Ground Bioengineering Techniques
for Slope Protection and Erosion Control

Also of interest

*Water Bioengineering Techniques
for Riverbank and Shore Stabilisation*
H.M. Schiechtl and R. Stern
0–632–04066–1

Ground Bioengineering Techniques

for Slope Protection and Erosion Control

H.M. Schiechtl

and

R. Stern

Translated by L. Jaklitsch
UK Editor David H. Barker

Blackwell
Science

© 1992 Österreichischer Agrarverlag,
Klosterneuburg, Austria
© 1996 English translation with additions
Blackwell Science Ltd

Editorial Offices:
Osney Mead, Oxford OX2 0EL
25 John Street, London WC1N 2BL
23 Ainslie Place, Edinburgh EH3 6AJ
238 Main Street, Cambridge
  Massachusetts 02142, USA
54 University Street, Carlton
  Victoria 3053, Australia

Other Editorial Offices:
Arnette Blackwell SA
  224, Boulevard Saint Germain
  75007 Paris, France

All rights reserved. No part of this publication may be reproduced, stored in a retrieval system, or transmitted, in any form or by any means, electronic, mechanical, photocopying, recording or otherwise, except as permitted by the UK Copyright, Designs and Patents Act 1988, without the prior permission of the publisher.

First published in German as *Handbuch für naturnahen Erdbau* by Österreichischer Agrarverlag, Austria
First published in English by
Blackwell Science 1996

Set in 10 on 13 pt Times
by DP Photosetting, Aylesbury, Bucks
Printed and bound in Great Britain
by Hartnolls Ltd., Bodmin, Cornwall

The Blackwell Science logo is a
trade mark of Blackwell Science Ltd,
registered at the United Kingdom
Trade Marks Registry

DISTRIBUTORS

Marston Book Services Ltd
PO Box 269
Abingdon
Oxon OX14 4YN
(*Orders:* Tel: 01235 465500
         Fax: 01235 465555)

USA
Blackwell Science, Inc.
238 Main Street
Cambridge, MA 02142
(*Orders:* Tel: 800 215-1000
              617 876-7000
         Fax: 617 492-5263)

Canada
Copp Clark, Ltd
2775 Matheson Blvd East
Mississauga, Ontario
Canada, L4W 4PZ
(*Orders:* Tel: 800 263-4374
              905 238-6074)

Australia
Blackwell Science Pty Ltd
54 University Street
Carlton, Victoria 3053
(*Orders:* Tel: 03 9347-0300
         Fax: 03 9349-3016)

A catalogue record for this title
is available from the British Library

ISBN 0–632–04061–0

Library of Congress
Cataloging-in-Publication Data

Schiechtl, Hugo M.
  [Handbuch für naturnahen Erdbau. English]
  Ground bioengineering techniques for slope
protection and erosion control/H.M. Schiechtl and
R. Stern.; translated by L. Jaklitsch.
    p. cm.
  Includes bibliographical references and index.
  ISBN 0-632-04061-0 (alk. paper)
  1. Soil stabilization. 2. Soil-binding plants.
3. Slopes (Soil mechanics) 4. Soil erosion.
I. Stern, R. II. Title.
TA749.S3513 1996
624.1'5363—dc20   96-5617
                                                  CIP

Contents

List of Colour Plates	viii
Preface	x
Editor's Note	xi
Introduction	xiii

1 Planning and Implementation — 1
  1.1 Aesthetic considerations — 1
  1.2 Minimising disturbance of the landscape — 2
  1.3 Planning of bioengineering protection works — 2

2 Ground Bioengineering Systems — 5
  2.1 Definitions — 5
  2.2 Function and effects — 5
  2.3 Biological building materials — 6
    2.3.1 Species selection — 7
    2.3.2 Vegetation systems and plant origin — 9
    2.3.3 Plant propagation — 10
  2.4 Preliminary works — 13
  2.5 Selection of the method and type of construction — 16
    2.5.1 Construction timing — 19
    2.5.2 Limits of application — 19
  2.6 Construction costs — 19

3 Ground Bioengineering Techniques for the Protection and Stabilisation of Earthworks — 47
  3.1 Soil protection techniques — 48
    3.1.1 Turfing — 48
    3.1.2 Grass seeding — 50
      3.1.2.1 Hayseed seeding — 52
      3.1.2.2 Standard seeding — 53
      3.1.2.3 Hydroseeding — 54

		3.1.2.4 Dry seeding	55
		3.1.2.5 Mulch seeding	55
	3.1.3	Direct seeding of shrubs and trees	58
	3.1.4	Erosion control nets	60
	3.1.5	Seed mats	60
	3.1.6	Precast concrete cellular blocks	62
	3.1.7	Live brush mats	63
3.2	Ground stabilising techniques		65
	3.2.1	Live cuttings	65
	3.2.2	Wattle fences	67
	3.2.3	Fascines	70
	3.2.4	Fascine drains	71
	3.2.5	Furrow planting	74
	3.2.6	Cordon construction (Praxl, 1961)	75
	3.2.7	Layering	77
		3.2.7.1 Hedge layers	77
		3.2.7.2 Brush layers	79
		3.2.7.3 Hedge-brush layers	82
	3.2.8	Gully control	84
	3.2.9	Stake fences	86
3.3	Combined construction techniques		87
	3.3.1	Vegetated dry stone block walls, stone pitching and revetments	87
	3.3.2	Filter wedge	88
	3.3.3	Vegetated gabions	90
	3.3.4	Vegetated geotextile earth structures	92
	3.3.5	Vegetated crib walls	94
	3.3.6	Live grating	97
3.4	Supplementary construction techniques		99
	3.4.1	Planting of root-ball container or pot plants	99
	3.4.2	Transplants	100
	3.4.3	Root divisions	101
	3.4.4	Transplanting rhizomes and chopped rhizomes	101
3.5	Special structures and techniques		102
	3.5.1	Rockfall protection	102
		3.5.1.1 Catch walls or barriers	102
		3.5.1.2 Suspended wire mesh	103
		3.5.1.3 Fixed protection nets	110
	3.5.2	Wind breaks or shelters	110
	3.5.3	Noise abatement structures	111

4 Care and Maintenance of Structures — 113
4.1 Fertilisation — 114
4.2 Irrigation — 115
4.3 Ground preparation — 115
4.4 Mulching — 115
4.5 Mowing — 115
4.6 Pruning — 116
4.7 Staking and tying — 116
4.8 Pest and disease control: prevention of browsing damage by wildlife — 116

Glossary — 121

References — 133

Further Reading — 135
Temperate and general ground bioengineering — 135
German language further reading — 139
Theses — 140
DIN standards and other codes of practice — 140

Index — 141

List of Colour Plates

Plate 1	Storage of stripped turfs
Plate 2 and 3	Cutting and loading of rolled turf in a nursery
Plate 4	Recently completed slope with placed turf
Plate 5	Slope stabilisation by turf in Alpine region – 50 years old
Plate 6	Grassed channels on highway cut slope
Plate 7	Hydroseeding with hydroseeder
Plate 8	Mulch seeding with long straw (the Schiechteln method). Work sequence: spreading of long straw, seeding and binding with a bitumen emulsion
Plate 9	Mulch seeding with long straw
Plate 10	Slope protection by straw mulch cover
Plate 11	Comparison with Plate 10 after six months
Plate 12	Placing of erosion control nets made of coir fibre and jute. (Courtesy Florineth, 1983)
Plate 13	Poor result of slope stabilisation: the pre-cast concrete cellular blocks are much too bulky to permit proper grassing
Plate 14	Slope planted with cuttings after one year
Plate 15	Poor take of a wattle fence due to desiccation
Plate 16	Fascine construction
Plate 17	Construction of fascine drains. (Courtesy Florineth, 1983)
Plate 18	Layout of furrow planting on a slope
Plate 19	Cordon stabilised major slide
Plate 20	Large-scale brush layer construction using a climbing back hoe
Plate 21	Manual construction of a brush layer
Plate 22	Embankment stabilisation by hedge–brush layer immediately after completion – Brenner Pass motorway, Austria
Plate 23	Comparison with Plate 22 after one year
Plate 24	Hedge–brush layer after three years: the vertical stems of the rooted plants can be clearly distinguished from the multi-stemmed growth of live brush cutting

Plate 25	Ten-year-old hedge–brush layer (compare with Plate 24)
Plate 26	Typical steep-sided gully suitable for brush layer construction
Plate 27	Completed layering
Plate 28	Dry stone block wall combined with live cuttings during construction
Plate 29	Dry stone block wall avalanche protection combined with cuttings and turf after ten years
Plate 30	Stone pitching combined with turf as an alternative to concrete walling
Plate 31	Filter wedge during construction
Plate 32	Gabions during construction
Plate 33	Gabions reinforced with willow cuttings after nine years
Plate 34	(*top left*) Geotextile reinforced slope with branch layers
Plate 35	(*top right*) Comparison with Plate 34 five months after completion: the willows are well established
Plate 36	Double stretcher log crib wall, combined with willow brush layers
Plate 37	Detail of Plate 36
Plate 38	Log crib wall after completion: the willows have not yet taken
Plate 39	Comparison with Plate 38 after five years
Plate 40	Concrete crib wall during construction: the willow brush layers are inserted at the same time
Plate 41	Concrete crib wall with willow brush layer after one year (compare with Plate 40): they reinforce and drain the fill
Plate 42	Construction of a simple wooden grating, using poles
Plate 43	Detail of a double grating using wooden poles
Plate 44	Double grating work for slope stabilisation
Plate 45	Comparison with Plate 44 after 15 years
Plate 46	Species composition plan for a 25-year-old stabilisation project by vegetative methods (hedge–brush layer) of an embankment of the Brenner motorway, Ahrnberg embankment; 700 m above sea level, facing west, on alluvial gravel. (Source: Schütz, 1989)
Plate 47	Slope stabilised by vegetative means before planting of woody species
Plate 48	The same slope shown in Plate 46, 15 years after planting of woody species
Plate 49	Transplantation of large turf panels complete with topsoil: small woody plants are contained in the panels

Preface

In recent times, there has been a profound change in people's attitude and awareness towards the environment. People from all walks of life now have a greater understanding of, and concern for, the deteriorating condition of their environment.

This has led to a greater emphasis on biotechnical methods in civil engineering, since these techniques offer one way of counteracting damage to the environment that is caused by major new industrial and housing developments and by new roads, with cuttings and embankments.

The main aim of this handbook is to bring vegetative or biotechnical methods of soil protection and slope stabilisation to the notice of as wide a range of people as possible engaged in practical work on site and to encourage greater use of bioengineering techniques. It also seeks to encourage close contact between the experts engaged in this field.

In order to apply these methods in the field, further training and information need to be available to those working on the practical aspects of erosion control and soil stabilisation. Unfortunately, the reliability of available mathematical slope stability models incorporating vegetative effects is hampered by the general lack of reliable input data.

It is hoped that this book, and its companion volume on *Water Bioengineering Techniques for Riverbank and Shore Stabilisation* will result in reduced rates of environmental degradation, enhanced safety and performance of engineering projects, and ultimately contribute to conservation of habitats and increased species diversity, wherever ground bioengineering is practised.

Editor's Note

This handbook, one of a pair, provides a rare opportunity to gain an insight into the approach of the chief exponent, Professor Hugo Schiechtl, and his colleague, Dr Roland Stern, of the use of vegetation for the engineering and ecological enhancement of earth structures. The book has been written after a lifetime engaged in developing for modern conditions the classical repertoire of earlier bioengineers in the Central Alpine region of Europe. The techniques described in detail in Chapter 3 have been built upon a profound knowledge of local soils, plants, their ecology, and slope processes. This has provided a foundation for protection and stabilisation of natural and formed slopes along transportation routes and adjacent to industrial and housing areas and leisure facilities.

The information in this handbook may not apply in all respects outside the region in which it was gathered. In addition to its scientific value, it provides a guide to the way botanical, ecological and geotechnical knowledge and construction know-how have been assembled, classified and co-ordinated for those engaged in bioengineering work in most regions of the world to emulate. The intentions of the authors in publishing the original work in German should be well-served by opening it up to the larger English-speaking world, for most of which the large body of German-language bioengineering literature is unfortunately closed.

There are many instances in the text where unspecified live cuttings are mentioned. In Europe, these will usually be willow species, and occasionally dogwood or poplar. Elsewhere, similarly vigorous vegetatively propagated indigenous shrub and tree species will be equally suitable.

This edition is based on the translation by Dr L. Jaklitsch of the German original text.

Synonyms

In general there are many different terms describing similar techniques involving the use of vegetation for civil engineering purposes in the four main active German-speaking countries or regions – Austria, Germany, Switzerland and the Alto Adige-South Tirol region in Northern Italy. A difficult problem is thus compounded for non-German speakers wishing to decipher original texts, increasing the value of this book.

As the practice of vegetative engineering has begun to spread around the world, the most commonly adopted English term to describe it is bioengineering, a direct translation of the most commonly used German-language term 'Ingenieurbiologie'. However, the disciplines of human and genetic engineering are both frequently called bioengineering and/or biotechnology. Since the medical and genetic use of the prefix 'bio' in conjunction with the words engineering or technical pre-dated has a much higher 'profile' than vegetative engineering topics, the shared use of the terms bioengineering without suitable prefix or 'identifying' link words can give rise to distracting confusion. This is made more complicated as genetic engineering of plants offers the prospect of improving the civil or geotechnical engineering performance of some plants to enhance their desirable characteristics.

Since the term 'soil bioengineering' has been defined by Sotir (1995) as involving plants exclusively, it is considered that the best prefix to use is 'ground', as in 'ground bioengineering'. This mirrors the associated term 'ground engineering' which is defined as 'concerning engineering processes for improving ground'. Ecological engineering or eco-engineering are other terms which have been proposed (Nordin, 1993; Sortir, 1995). Also 'biotechnical slope protection' is used frequently by bioengineering practitioners in the USA to describe the combined use of structural and vegetative elements to arrest and prevent slope failures and erosion (Gray, 1991).

Acknowledgements

The assistance of Dr Neil G. Bayfield on ecological aspects of the text and Mrs Klare Ware on the translation from the German of glossary items and communications with the authors is gratefully acknowledged. The glossary and English bioengineering references have been compiled from the original text and unpublished documents prepared by Geostructures Consulting. Thanks also go to Julia Burden at Blackwell Science for her faith in the text, the discipline of ground bioengineering and the editor.

David H. Barker
Managing Director, Geostructures Consulting, Edenbridge

Introduction

Since time immemorial humankind has had to safeguard its living space and the land used to provide for its needs against the forces of nature. The earliest settled communities achieved this by the most simple means, the maintenance of the natural environment.

Even in those early times, ways and means had to be found to stabilise and protect those parts of the landscape which had been artificially altered. This was achieved by the use of local materials such as stone, wood and plants – the first application of ground bioengineering techniques in the widest sense of the word.

At the beginning of the nineteenth century, experience was gained in repairing damage caused by natural catastrophes, and during that century increased systematic research in the areas of geology, geography and geobotany augmented the knowledge of the distribution and habitat requirement of plants. During the twentieth century refined monitoring methods paved the way for improved system analysis, thus providing more knowledge about the eco-physiology of plants and animals and about the mechanical properties of plant materials.

It was only in the 1950s that researchers in various disciplines started more intensive and systematic work on the applicability, effectiveness and ultimate potential of vegetative methods in the field of civil engineering. This started a process in which the use of plants and plant materials for the protection and stabilisation of earthworks was developed. To describe these systems, Kruedener (1951) employed for the first time the term 'Ingenieurbiologie', or in English 'bioengineering techniques'. Today, these methods comprise an inter-disciplinary subject which in the face of rejection, sceptical acceptance, or exaggerated expectations, as the case may be, must strive to assert its rightful place within the spheres of technology and biology.

Chapter 1
Planning and Implementation

Earthworks are a necessary part of many civil engineering projects, for example road construction, river works, avalanche protection, the construction of sports facilities, skiing pistes, rock dumps resulting from tunnel construction and mining operations, landfills, and so forth.

There is no doubt that large-scale earth movement leaves its mark on the landscape, leading to change in its appearance and, to a lesser degree, its local ecology. This change does not necessarily have to be totally negative. Several guidelines and suggestions with regard to project planning and relevant construction methods are outlined: these are designed to lessen the negative impacts and to preserve as far as possible the appearance of the natural environment.

1.1 Aesthetic considerations

The rapid and cost-effective implementation of the project should not be the sole consideration of the project plan: for eco-environmental and aesthetic reasons, the project structures should fit as harmoniously as possible into the landscape, and after the completion of the works, it should be restored as completely as is feasible to its original appearance. An important condition for this objective to be achieved is the balanced distribution of cut and fill.

The use of local building materials or their nearest equivalent should be given preference in the construction process. Extremely deep cuts or cutting into very steep slopes and excessively high embankments should, wherever possible, be avoided. Ecologically valuable areas and landscapes of great beauty should not be unnecessarily fragmented. In particular, forest and woodland edges, coppices, spinneys and wetlands, etc. should receive due consideration. The necessary biotechnical aspects of the project should be covered in the project report and dealt with more specifically in the relevant sections of the detailed plan.

1.2 Minimising disturbance of the landscape

Some important aspects of landscape-aware construction methods are as follows:

- Selection of suitable construction machinery and tools: they should be technically suitable, but also suitably sized to fit the task. For example, the use of excavators and similar equipment for the construction of haulage roads in forests.
- Limited removal of solid rock, avoiding scattering of rock by the use of restricted blasting techniques. Partly decomposed and well-jointed rock is often rippable using tines on heavy crawler equipment, or jack hammers.
- Excess material should not be dumped downslope, but removed to predetermined dump sites. Timely provision should be made of ancillary protection works such as retaining walls, wattle fences or similar barriers to safeguard against rockfall and soil loss down slope.
- Adequate drainage to prevent subsequent damage from excess run-off and to reduce maintenance costs.
- Stable embankments to fit the terrain, avoidance of steep side-slopes and timely provision of protection structures preferably using vegetative methods.
- Conservation of any woodland, spinney, etc. for noise protection and appearance.
- In closed forests, completely cleared areas must be kept to an absolute minimum.

1.3 Planning of bioengineering protection works

There is still a marked reluctance to use vegetative methods in preference to the well-established practices employed in conventional civil engineering. This is most likely due to a lack of training or lack of personal experience in a relatively new field.

It is a common occurrence for vegetative methods not to be considered at the preliminary planning stage of a project, but only resorted to when the generally accepted methods have failed. They are then, more often than not, hastily planned, ill-prepared and implemented under pressure of time.

It is therefore of value to consult, at the preliminary project planning stage, an experienced engineer and/or landscape planner to explore the feasibility of integrating the bioengineering aspects and measures into the overall plan of civil engineering construction.

A certain preparation time is, however, required to find the most suitable and economic solution. The final decision concerning the choice of the individual construction methods, together with the necessary plant material, can only be made after close familiarisation with the construction site has been achieved and the wishes of the client carefully noted.

With the aid of the checklist (Table 1.1), it can be ascertained what information and which facts are absolutely necessary at the planning stage. The decision as to which vegetative method to adopt and the selection of the required plant material will ultimately dictate the optimum time schedule for the construction stage.

Table 1.1 Checklist for the preparation of a bioengineering project.

No.	Type of work
1	Obtain topographical surveys and maps, aerial photos, orthophotos* and working drawings
2	Evaluation of geological and hydro-geological investigations
3	Evaluation of the pedological investigation (soil types)
4	Vegetation survey and mapping of the area surrounding the construction site
5	Collection of ecological data (site investigation)
6	Investigation of the cause of damage in the case of repair works
7	Determination of the target to be achieved by the proposed measures
8	Selection of type of plant and building materials to be used
9	Selection of construction types and methods to be used
10	Legal aspects (ownership and land use rights, indemnities, etc.)

*Orthophotos are aerial photographs corrected for distortion to conform to map accuracy at a given scale.

Chapter 2
Ground Bioengineering Systems

2.1 Definitions (Kruedener, 1951)

Bioengineering techniques are based on harnessing biological and, in particular, botanical knowledge in implementing protection and stabilisation measures for earthworks, hydraulic structures, river banks, gullies and other features in the formed and natural landscape. Their characteristic features are the use of plants and plant materials which, when employed either on their own or in conjunction with inert building materials, will bring about the stabilisation of earthworks. They are not seen as a substitute for, but as a necessary and logical extension and complement to accepted and proven solely mechanical civil engineering methods.

2.2 Function and effects

It is normally inevitable, even with the most careful planning, that civil engineering projects interfere with the natural terrain, thereby modifying the landscape. They involve the excavation and movement of large volumes of soil, which in turn may lead to the well-known phenomena of erosion, rock falls, landslides, etc. The shaping and protection of newly-established surfaces and areas is therefore of great importance. Vegetative methods are to a large extent eminently suitable for this purpose, as their efficacy is not limited to technical aspects; they also effect aesthetic, ecological and economic benefits. Depending upon the type and manner of construction chosen, different effects may be achieved (Table 2.1). The late adoption of ground bioengineering methods as remedial measures to rectify or even merely to ameliorate any shortcomings in the tender specifications, or flaws in the construction of earthworks, is rarely, if ever, fully effective.

Table 2.1 Multiple effects of ground bioengineering systems.

Geotechnical	• Protection of the soil surface from erosion caused by precipitation, wind and frost.
Ecological	• Moderation of temperature and moisture extremes of the air at ground level, creation of near ideal growing conditions in the vegetation zone. • Improvement in the soil water status (drainage and retention) by way of water interception, evapotranspiration, and increased water capacity. • Soil improvement and humus formation from decaying and decomposing vegetation resulting in a build-up of soil fauna and flora and subsequent increase in nutrient content. • Creation of new and better living conditions for plants and animals.
Economic	• Construction and maintenance lower costs. • Creation of usable areas for agriculture, forestry, housing and industry.
Aesthetic	• Integration of buildings and construction elements into the landscape, rendering it more attractive.

2.3 Biological building materials

The following is a summary of some of the plants and plant materials used:

- seeds of grasses, herbs, trees and shrubs;
- parts of woody plants capable of vegetative propagation to be used as cuttings, stakes, poles, branch layers, root divisions, etc;
- parts of herbs and grasses suitable for vegetative propagation, such as rhizomes, stolons, slips, etc;
- saplings and rooted shrubs, etc;
- turf, sods, complete with topsoil and vegetative cover of young trees, shrubs, herbs and plant associations.

The quality of the live plant materials used must meet certain standards and specifications, particularly with regard to origin, age, size and absence of diseased parts. In principle, best results are achieved when it is possible to use indigenous plants and plant material. *In situ* vegetation, which is capable of being re-used, should be carefully removed before construction starts, stored and reinstated after the completion of the works. Trees and shrubs capable of vegetative propagation should be obtained from the same area for later use.

The selection of the correct plant material is essential for the permanent success of bioengineering protection measures. The selection and use of unsuitable plant material usually leads to failure.

2.3.1 Species selection

Four broadly-based criteria, individually evaluated or used in combination, should be applied in the process of selecting suitable species:

- ultimate objective of the measures to be taken;
- ecological make-up of the species;
- biotechnical make-up of the species;
- origin (or provenance).

The goal of measures taken is primarily the stabilisation of slopes. Together with the aim of obtaining the desired effective stabilisation, the choice of plants should also aim at the establishment of easily maintained and, preferably, economically valuable grassland and/or woodland/shrub formations.

The ecological character of vegetation reflects its reaction to the conditions of the habitat. Plants and plant associations are site indicators and reflect the prevailing conditions of the local habitat.

For best results, only plants that grow in similar habitats should be used; the simplest and best-suited method for plant selection should be based on the results of plant sociological surveys of nearby similar habitats. Most vegetative stabilisation works are carried out on the slopes of artificially created cuts and fills, devoid of topsoil, i.e. on relatively inert subsoils or decomposed rock of very poor fertility.

Only well-adapted pioneer plant associations with a large spectrum of species, each tolerant of a wide range of adverse factors with regard to soil quality, micro-climate and mechanical stress, have any chance of establishing themselves under such rigorous conditions. Plant species with a wide amplitude of ecological versatility include the following:

Trees	Grey alder (*Alnus incana*)
	European larch (*Larix decidua*)
	False acacia (*Robinia pseudacacia*)
	Sallow (*Salix caprea*)
	Silver birch (*Betula pendula*)
	Black poplar (*Populus nigra*)
	Scots pine (*Pinus sylvestris*)
Shrubs	Dogwood (*Cornus sanguinea*)
	Hoary willow (*Salix eleagnos*)
	Fly honeysuckle (*Lonicera xylosteum*)
	Privet (*Ligustrum vulgare*)
	Purple osier (*Salix purpurea*)

	Elder (*Sambucus nigra*)
	Black willow (*Salix nigricans*)
Grasses and legumes	Creeping bent (*Agrostis stolonifera*)
	White melilot (*Melilotus albus*)
	Perennial ryegrass (*Lolium perenne*)
	Bird's-foot trefoil (*Lotus corniculatus*)
	Cocksfoot (*Dactylis glomerata*)
	Red clover (*Trifolium pratense*)
	Red fescue (*Festuca rubra*)
	Sweet vernal grass (*anthoxanthum odoratum*)
	White clover (*Trifolium repens*)
	Smooth meadowgrass (*Poa pratensis*)
	Kidney vetch (*Anthyllis vulneraria*)

The technical term 'biotechnical character' of a plant encompasses a number of features. Plants of high biotechnical value should possess the following attributes:

(1) The ability to take root in, and colonise immature soil, subsoil or indeed any kind of material exposed in cuts or used as fill – this ability characterises pioneer plants which, by taking root on bare ground, prepare the soil for further plant succession. They are predominantly species with a broad ecological amplitude.

(2) Roots and surface parts must be resistant to the mechanical forces of erosion and aggradation, the forces exerted by snow cover, rock fall, as well as subsoil movement. Internal soil movement (friction and shear) must be specially considered for slope stabilisation measures and combined construction methods.

(3) Soil strengthening or soil binding effect – this depends on the type of roots, the intensity of root penetration and on the total root mass. Extensively rooted plants with a spreading and/or deep root system (trees, shrubs) develop horizontal or vertical root patterns. Mature trees and shrubs propagated from seedlings are divided into shallow-rooted or deep-rooted groups. *In situ* vegetatively propagated trees and shrubs cannot be placed with confidence in any of these groups. The group with intensive, rather shallow, but much-branched root systems that form a dense mass is represented by the grasses. Among the herbaceous plants – particularly those belonging to the legumes – are transitional types which cannot be fitted into either group. The most effective soil strengthening is achieved when root penetration extends into the lower subsoil. For best results, it is therefore essential that several different plant species are used in combination.

(4) Soil improvement – this concerns the ability of the planted vegetation to improve the soil quality, which in turn leads without further intervention to the natural progression from the pioneer stage to the next higher plant association. This effect is achieved through soil strengthening and improvement of the micro-climate. Vigorous growth and the ensuing increase in plant material play an important role. Of particular benefit are plants which are able to improve the nitrogen status of the soil through root nodules and/or leaf mulch, in the first instance, legumes and alders.

(5) Salt tolerance – important in areas where high application rates of salt extend over long periods of time for reasons of traffic safety by reducing ice formation. Trials in the Inn Valley in Tyrol by Schiechtl (1983) have shown that only few woody plants can cope with these conditions. No species is completely resistant, but some are able to recover even after repeatedly suffering damage over several years, enabling them to fulfil their function as anti-dazzle centre strips. The following species fall into this category:

- Snowberry (*Symphoricarpos racemosus*)
- Elaeagnus (*Elaeagnus angustifolia*)
- Caragana (*Caragana arborescens*)
- Fly honeysuckle (*Lonicera xylosteum*)
- Japanese honeysuckle (*Lonicera tatarica*)
- Common ash (*Fraxinus excelsior*)
- Gooseberry (*Ribes alpinum, Ribes uva-crispa*)
- Rose (*Rosa rugosa*)
- Duke of Argyll's tea-tree (*Lycium barbarum*).

2.3.2 Vegetation systems and plant origin

The potential natural distribution of vegetation in Mainland Europe is separated into regional vegetation zones (Meusel *et al.*, 1965) or vegetation areas (Tschermak, 1961). In the present context, the broader divisions comprise the Central Alpine, the more Continental Region, the Atlantic (Oceanic) influenced Northern, and the Mediterranean influenced Southern fringe of the Alps. Within these floristic zones, elevation above sea level determines the occurrence and vigour of plant life. The selection of plant materials for construction purposes must therefore take due cognisance of the vegetation zone or area and its elevation above sea level.

In principle, the aim should be to use only plants and trees from vegetation areas which are as close as possible to the construction site. The higher the elevation of the site, the more vital it is to observe this basic rule.

2.3.3 Plant propagation

Consequent to the important principle of plant provenance and the necessity to have large quantities of plant material at one's disposal, economic considerations demand the easy propagation of the plants involved to prevent bottlenecks and shortages during construction. The propagation of biological building materials is either generative, using seed, or vegetative, by the use of shooting or rooting plant parts. It would be of great advantage to the engineer using vegetative methods to have recourse to standard seed mixtures.

Unfortunately, it is not possible to obtain commercially available grass seed mixes which are made up of those species that constitute the natural succession for any given site. Seed collection from indigenous grasses growing in the wild has ceased a long time ago as it is economically not viable: it is resorted to only on rare occasions for very special projects. Standard hayseed, collected from haystacks and barns in the vicinity of the construction site, is therefore a valuable and sought-after resource. Site conditions are, however, not the only criterion for the choice of mixes: cost factors also merit careful consideration.

Most of the grass seed freely available from the seed trade is bred for agricultural use and for ornamental lawns; it is, on average, not adapted to the harsh conditions that prevail at the construction site. Single variety seed is, as a rule, only of limited value for earth protection works. Multi-species seed mixtures are more natural and invariably more stable.

Important criteria for the selection of a suitable mix are the expected pressures which the established cover has to withstand, and the expected life span. The grass cover of a skiing piste, for example, is subjected to much greater stress than that of the average cut or embankment slope. The grasses sown on the piste must form stable associations of a permanent nature which can cope with the tremendous pressure occurring during the skiing season.

Practical experience has shown that a careful selection of the most suitable seeds of grasses, herbs and woody plants from local nurseries and seed merchants is usually satisfactory.

Several countries recommend for certain regions, or even for isolated local habitats, so-called standard seed mixes and seed merchants offer such mixtures to the user, blended according to their experience. For large area plantings and for extreme, difficult locations, it is recommended that seed mixes be prepared on the advice of experts, or by special seed breeders. Seed obtained from normal outlets should be tested in specialist laboratories with regard to suitability and quality. The relatively small cost for this is worthwhile.

Recipes for seed mixtures of woody plants require great experience. Apart from the seeds of woody species, such mixtures should contain as a rule seeds of grasses and herbs, because the actual seeding does not just aim at the establishment of some kind of shrubbery or woodland, rather is its primary objective the rapid stabilisation of the slope. Species of grasses and herbs with deep rooting characteristics or with smothering top growth habit, which on establishment would lead to serious competition with the woody plants, are, for obvious reasons, to be excluded from the mix.

Table 2.2 at the end of this chapter lists the properties of some of the more important seeds and plants that are suitable for the vegetative methods of soil protection. The most important live materials for slope stabilisation and combined construction measures are parts of woody plants with adventitious buds. Such plant parts capable of vegetative propagation are separated into various size groups:

- Cuttings: unbranched stems, approximately 12–60 mm in diameter, 250–600 mm long.
- Branches and twigs: should be at least 600 mm long and of various thicknesses.
- Stems: slim, flexible, poorly branched shoots with a minimum length of 1200 mm.
- Poles or truncheons and stakes: straight, mainly unbranched sections of large branches or boughs 1–2.5 m long.

Ideally, fairly thick and long branches should be used, suited to the construction method envisaged, as the more substantial cuttings will ensure successful sprouting and rooting. Experience has shown that the best results are achieved by the use of twigs and branches varying from finger thickness up to about 80 mm in diameter. Thin stems and twigs dry out easily and are best used only in conjunction with more substantial plant parts. The required amounts of plant material may be procured from the following sources:

- Branches and other parts may be obtained from nearby ecologically similar woodland.

- Thinning and pruning of established nearby protection works, planted with suitable species, may provide adequate quantities.
- If necessary, the nearest nursery should be approached, if natural sources as mentioned above are not available.

The optimum time for obtaining cuttings and poles is during the period between leaf fall and budding, i.e. October to April.

Shrubs and young trees are cut just above ground level; older trees have the branches lopped off in a balanced and systemic manner. To obtain a smooth and clean cut, it is best to use secateurs or a saw. In order to protect the branches from desiccation, they are transported whole to the construction site and should be cut into the required section just before use. In principle, the aim should be the immediate use of the plant material, taking particular care to ensure that the branches, stakes and poles are well embedded in the soil to prevent drying out.

If immediate planting is not feasible, cuttings obtained during plant dormancy may be preserved for reasonable lengths of time if they are protected against drying and bacterial heating. This may be achieved by storing the parts in snow, submersion in flowing water of maximum 15°C or storage in cold rooms of 0–1°C in PVC bags, or wrapped in PVC film. Anti-transpiration sprays can prevent desiccation. Plant material which has thrown shoots cannot be stored for future use.

Approximately 30 species of trees and bushes suitable for vegetative propagation, mainly willows, are available in Central Europe. Of these, some ten willow species are specific to the montane and sub-alpine region of the Alps (Schiechtl, 1980; 1986; 1992). Woody species capable of producing shoots and their properties are detailed in Table 2.3 at the end of this chapter. All woody plants that can be vegetatively propagated (Table 2.2) may, of course, be planted as rooted seedlings or saplings.

Rooted seedlings of non-vegetatively propagating woody plants are used on their own in the 'hedge layer method' (see Section 3.2.7.1), or in combination with branches in the 'hedge-brush layer method' (see Section 3.2.7.3). Well-suited for this purpose are pioneer plants which can survive complete accidental soil burial and are able to produce adventitious roots. Conifers are not suitable.

The most important trees and shrubs suitable for use in vegetative methods of slope stabilisation are listed in Table 2.4 at the end of this chapter.

Most trees and bushes occurring in any given vegetation zone are, within their area, suitable for horticultural purposes and landscape gardening. However, only a few are suitable for vegetative methods of soil protection and slope stabilisation. It is primarily those capable of vegetative propagation and the species listed in Table 2.3 that are particularly suited for the latter purpose. Those species which are not primarily suited may be introduced at a later stage after slope stabilisation has been achieved.

2.4 Preliminary works

The purpose of preliminary works is the protection of the construction site and the safety of the workforce. This may be achieved through the use of temporary barriers or fences made of inert 'conventional' building materials. They are to prevent or stop landslides, rock falls and downward movement of debris, soil or other material in general.

Temporary preventive drainage systems of 'hard' or rigid construction will at a later stage be replaced by fixed drains, employing, if required, vegetative or combined methods of construction (see Sections 3.3–3.5). If serious local erosion has damaged the vegetation cover and land surface, certain corrective measures must be taken before vegetative protection measures are contemplated. Ridges are to be levelled. Of particular importance is the shaping of overhanging, under-washed eroded crests of slopes. Such crests are very prone to continuous erosion and soil movement, and need to be carefully rounded off in order to be stabilised. This is effected by cutting and removing the overhanging section of the erosion rim, thus establishing a transition arc of a minimum radius of 5 m (Figs 2.1 and 2.2). The flattening of excessive slopes may be achieved by various means, e.g. by hand, hydraulic or mechanical means.

- Loose or easily broken up soil may be removed by the use of hand tools such as hoes, picks, spades and shovels. Non-rippable and rocky conditions require the use of appropriate machinery and implements, or blasting.
- The removal of soil hydraulically relies on the energy of a high-pressure jet. This method offers considerable advantages for the regrading of steep slopes at inaccessible construction sites in mountainous terrain; it is therefore to be used in preference to mechanical means. Water is piped to the site from a provisional storage work or container situated on higher ground using quick coupling or PVC pipes, ending in a high-pressure nozzle. If it can be done, and there is no danger to the work team, the jet should be aimed from directly below, as the undercutting action of the jet causes large earth masses to fall. All washed off material accumulates at the lower slope or slope base, saturated with water, as in the course of natural sedi-

mentation. Silt fences should be erected to trap fine sediments on site to prevent pollution of adjacent soils and drainage systems.
- As a rule, soil removal is achieved by earth moving machinery, which is selected to suit the prevailing ground conditions.

Only subsoil should be removed by the use of a dozer blade. Where natural topsoil is still present, all shaping work is best done by the use of mechanical buckets, excavators or back hoes because they are more

Fig. 2.1 An unsufficiently rounded off slope crest is a continuous source of trouble.

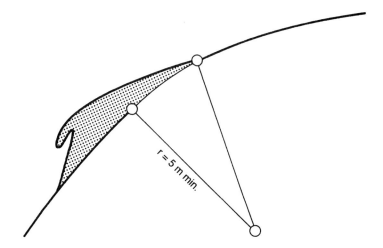

Fig. 2.2 Rounding off of crests of slopes.

efficient at removing the valuable and fertile topsoil layers for later reinstatement. The more extreme and sensitive the terrain, the lighter the equipment to be used on site. Articulated or 'climbing' backhoes, (e.g. Menzi Muck (Switzerland) and Kamo (Japan)), have been used to great advantage on steep slopes and at work sites without proper road access. Their use causes the least damage to the vegetation.

All material obtained from the flattening of the slopes must be deposited on specially prepared dump sites below the slope or directly on the slope. Transporting the soil to preselected tips further away should be considered. This allows not only for the safe deposition of the material, but also it could be used to advantage for terrain improvement. Such dumps are finally planted to grass and/or stabilised by other vegetative means.

Establishing smooth, excessively even and large slope areas is not of advantage, as this complicates revegetation of the surface. Moreover, smooth slopes of uniform inclination are perceived as unnatural in the landscape. For this reason, only rough levelling should take place, removing large stones and boulders only if rock falls are anticipated. Breaking a uniform slope by berms or terraces should only be attempted if some technical or economical advantage is thereby obtained, such as, for example, effective drainage, prevention of rockfalls, snow dumps and the provision of vehicular access for planting and maintenance operations (Plate 25). After completion of the terrain correction measures, the previously removed rooted topsoil, together with its vegetation cover, is replaced and suitably stabilised by vegetative means. The costs for such shaping operations often exceed those of the stabilisation works; nevertheless, they are unavoidable to ensure ultimate success.

Attempts at cost cutting in this respect are ill-advised and may jeopardise the whole operation. Under certain circumstances, it may be necessary to improve the quality of infertile or even sterile soils to create suitable conditions for the establishment of a vigorous vegetative groundcover.

One of the most effective means to achieve this is restoration to its former location of good quality topsoil which was removed during the course of the building operations, or spreading of topsoil collected from areas other than the construction site. Soil formation is a long-term process, often taking centuries. In areas with extreme growing conditions, even small amounts of topsoil are of great importance. On gently undulating or level terrain, the depth of redistributed topsoil should never exceed 100 mm; on moderate slopes, this should not exceed 50–100 mm. On very steep slopes, no topsoil should be applied at all as it is likely to slide or be washed off. If at all possible, topsoil should be rotovated into the surface of the slope to avoid abrupt soil transitions. Compost is a particularly effective soil ameliorant and its application is essential for the successful planting of shrubs and trees under adverse conditions. Composts offered for sale by nurseries and similar outlets are of very variable quality. Many composts derived from rubbish show either high heavy metal concentrations, or they are not sufficiently decomposed, which often leads to heavy infestation with unwanted weeds. If chopped bark is to be applied as a soil improver, it is important that the bark particles do not exceed a size of 10 mm before being composted.

2.5 Selection of the method and type of construction

The choice of method and type of construction has to take into account the criteria for species selection. In addition, the project schedules must allow for sufficient time during that season which has been identified as the optimum period for the selected type of work.

A summary of the conditions that decide the choice is as follows:

- Objective of the works: the immediate aim is the stabilisation of the slope. In second place is the need to shorten the time required for natural plant succession, followed by the lowest possible maintenance costs and the creation of economically usable areas in the widest sense of the word.

- Technical effects: the protection of erodible slopes by means of an effective plant surface cover. Slopes prone to slides are protected by ground stabilising techniques or combined construction methods.
- Ecology of the site: site conditions influence the choice of plants, which in turn determines, amongst other things, the type of construction.
- Availability of relevant building materials.
- Time of year: methods that require the use of vegetatively propagated trees and/or bushes must be carried out during late autumn or in winter.

New ideas in landscape architecture are often expressed, and supplementary construction methods can be employed to translate them into reality.

The achievements of ground bioengineering construction techniques have for some time now exceeded the basic concept of protection and stabilisation, revegetation, ground cover, aesthetics, and enhanced ecological merit. After the ecological evaluation of a construction site, other factors, biotic and abiotic, such as potential damage caused by man, insects, fungus infection, wildlife and grazing animals, fire hazard and potential deleterious effects caused by pollution, should receive due consideration. As bioengineering methods are, in the first instance, mainly concerned with aspects of safety and failure preventative measures, a thorough knowledge of the engineering properties of the soil and ground engineering or geotechnical aspects is of great advantage.

The stabilising effect of vegetative protection works is influenced by a series of different parameters:

- type of works (cutting, embankment, etc.);
- slope dimensions (height, width, length or inclination, total area);
- slope shape (uniform, convex, concave, stepped);
- soil type (cohesive or non-cohesive);
- soil–water relationships;
- rooting characteristics of the established vegetative cover (type, shape, distribution).

The overall view as shown in Fig. 2.3 is a starting point for this process. It facilitates selection of the appropriate category of bioengineered construction methods for a range of specific slope/embankment stabilities for the matrix of strategies.

18 Ground Bioengineering Techniques

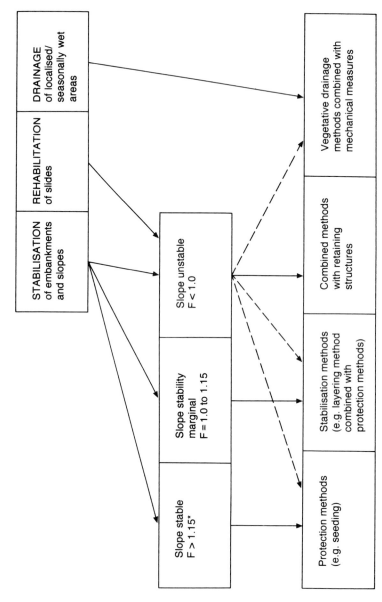

Fig. 2.3 Slope stability and bioengineering techniques. (Source: Kuonen, 1983).

Plate 1
Storage of stripped turfs

Plate 2 and 3
Cutting and loading of rolled turf in a nursery

Plate 4
Recently completed slope with placed turf

Plate 5
Slope stabilisation by turf in Alpine region – 50 years old

Plate 6
Grassed channels on highway cut slope

Plate 7
Hydroseeding with hydroseeder

Plate 8
Mulch seeding with long straw (the Schiechteln method). Work sequence: spreading of long straw, seeding and binding with a bitumen emulsion

Plate 9
Mulch seeding with long straw

Plate 10
Slope protection by straw mulch cover

Plate 11
Comparison with Plate 10 after six months

Plate 12
Placing of erosion control nets made of coir fibre and jute. (Courtesy Florineth, 1983)

Plate 13
Poor result of slope stabilisation: the pre-cast concrete cellular blocks are much too bulky to permit proper grassing

Plate 14
Slope planted with cuttings after one year

Plate 15
Poor take of a wattle fence due to desiccation

Plate 16
Fascine construction

Plate 17
Construction of fascine drains. (Courtesy Florineth, 1983)

Plate 18
Layout of furrow planting on a slope

Plate 19
Cordon stabilised major slide

Plate 20
Large-scale brush layer construction using a climbing back hoe

Plate 21
Manual construction of a brush layer

Plate 22
Embankment stabilisation by hedge–brush layer immediately after completion – Brenner Pass motorway, Austria

Plate 23
Comparison with Plate 22 after one year

Plate 24
Hedge–brush layer after three years: the vertical stems of the rooted plants can be clearly distinguished from the multi-stemmed growth of live brush cutting

Plate 25
Ten-year-old hedge–brush layer (compare with Plate 24)

Plate 26
Typical steep-sided gully suitable for brush layer construction

Plate 27
Completed layering

Plate 28
Dry stone block wall combined with live cuttings during construction

Plate 29
Dry stone block wall avalanche protection combined with cuttings and turf after ten years

Plate 30
Stone pitching combined with turf as an alternative to concrete walling

2.5.1 Construction timing

The most suitable time of the year during which vegetative methods of construction can be carried out is, to a certain degree, determined by the growing cycle of the plants which in turn is dependent on, and governed by, seasonal factors. All construction work which uses vegetatively propagated plant material must take place during the vegetative dormancy period (October/November to March/April). Grass seeding takes place during the periods of active growth; seeding of trees and shrubs, however, during spring and autumn. Rooted woody plants are best planted during spring or autumn; that is, at the beginning or at the end of the growing period. Potted or container plants may be planted throughout the summer period (Fig. 2.4).

2.5.2 Limits of application

The use of live building materials is limited by biological, technical and timing factors:

- Biological limits: areas which preclude the growing of tall woody plants. Observation of the limits of distribution; trees, for example, cannot be used in the higher Alpine regions.
- Technical limits: slope reinforcing is only possible in material which is capable of being penetrated by roots. Subsurface soil movements can only be indirectly eliminated by vegetative methods and subsurface drainage by evapotranspiration from a dense vegetation cover and root reinforcement of soil.
- Time limits: work during both growing and dormancy periods. Due to the limited application possibilities of bioengineering, it is obvious that live building materials cannot be employed to replace technical engineering methods, but only to supplement them.

2.6 Construction costs

The construction costs of the various methods are taken directly from practical experience gained in the field.

In order to eliminate the fluctuations of the value of money, construction costs are expressed in work hours. For direct cost comparison of the individual construction methods see Fig. 2.5. Costs, of course, vary with location, availability of construction materials and company

20   Ground Bioengineering Techniques

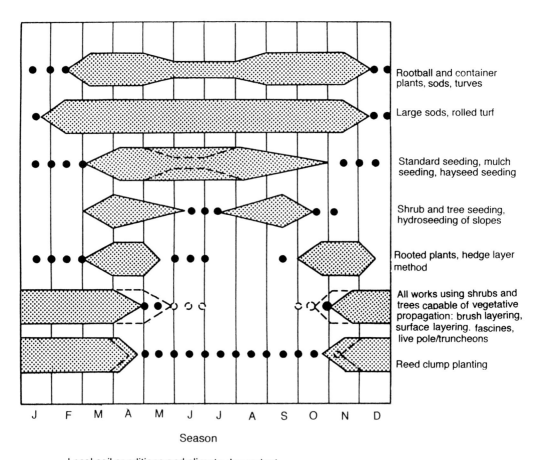

Local soil conditions and climate dependent.
UK variations are shown by dotted lines and circles.

Fig. 2.4 Installation schedule for bioengineering techniques, valid for Western and Central Europe, including Alpine regions, and other temperate zones of the Northern hemisphere.

structure. Accurate costings are therefore only possible when firm tenders are at hand. The relatively low cost of vegetative methods as compared with those of 'classical' civil engineering is often cited as the essential advantage of the former over the latter. This is not always the case, and occasionally vegetative methods may be more expensive.

Nevertheless, it is often possible to save by using vegetative methods, but a different approach is required:

❏   Masonry or stone walls may be constructed to reduced size or dispensed with altogether if they are replaced by more voluminous

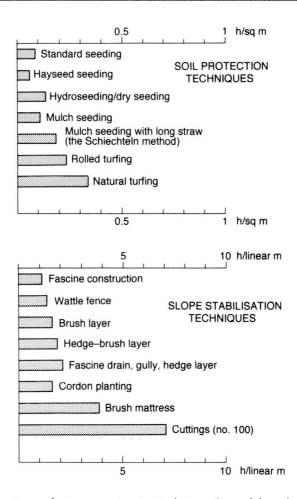

Fig. 2.5 Cost comparisons of various construction techniques (in work hours).

earthworks which are subsequently stabilised by vegetative methods. Massive walls may be replaced by stone pitching, suitably combined with live plant material. Gravel filters, planted to grass and more deeply rooted woody species, may in certain instances replace walls.

❑ Gully control measures and riverbank protection works are advantageously constructed of live material, or a combination of plants and stone.

❑ By the establishment of a good vegetative cover on all suitable surfaces immediately after the completion of the earthworks, stability is achieved before erosion occurs. Experience has shown that vegetative protection methods in erosion-prone terrain are on average not more expensive than the cost of remedial measures necessary after heavy rain or melting snow has damaged slopes not protected by such methods.

❑ Immediate vegetative protection of slopes permits steeper gradients. The angle of repose of fill is invariably steeper than the angle of the formed slope. The difference between the two is sometimes so great that it is possible to use the savings gained to finance the planting of grass and shrubs. This should, however, never tempt the designer to incorporate excessively steep slopes which, in spite of a good vegetative cover, may never be stable.

Much work has been done in various countries to ascertain the cost of vegetative methods expressed as a proportion of total project costs. It is of interest to note that these costs are subject to large variations depending on the amount of expensive 'hard' construction work involved.

As a rule, the proportional costs for vegetative protection works increase if the landscape is to be restored as closely as possible to its original appearance; at the same time total projects costs, however, decrease. For a range of applications, the proportional costs of vegetative construction methods amount to:

❑ road construction: 0.3–16%;
❑ power station projects: 0.2–2%;
❑ winter sports facilities: between 1 and 15% of total costs.

Annual maintenance costs are considerably lower than those required for conventional civil engineering structures.

It is a special feature of vegetative construction methods that maintenance and upkeep operations during the first few years after project completion are particularly important in order to ensure success. The type and duration of this maintenance period should be adequately covered in the tender specifications. When the desired vegetation cover has been fully established, and the objective has thus been achieved, natural development will take over, obviating further input apart from periodic care and maintenance – usually every few years. (See Chapter 4.)

The proper selection of ecologically suitable plant material together with appropriate construction methods will ensure that subsequent maintenance operations and costs are kept to a minimum.

Table 2.2 Seeds of grasses, herbs and woody plants suitable for vegetative methods of construction. (Source: Ehrendorfer, 1973.)

Latin and English name	Plant association*, habitat and properties	Morphology: stem and root	1000 grain wt	Seeds/g
Grasses				
Agropyron repens (Couch grass)	*Agropyro–Rumicion, Artemisietea, Chenopodietea, Festuco–Brometea, var. glaucum* in *Mesobromion* and *Erico–Pinion*. Arable weed, not to be used near arable land, needs light. Dry to moist, fertile; up to 900 m a.s.l., perennial.	Stem 0.2–1.5 m Root up to 0.8 m rhizomatous	4	260
Agrostis canina (Bent grass)	*Carici–Agrostidetum.* Fens and bogs, peat areas, wet woodland, pioneer plant of open wet soil. *Subsp. canina* on moist, *subsp. montana* on dry habitat of poor fertility; up to 1100 m a.s.l., perennial.	Stem 0.2–0.6 m Root 0.2 m	0.05	20000
Agrostis gigantea (Black bent, red top)	*Calthion, Glycerion, Caricetum fuscae, Molinio–Arrhenatheretea, Phragmitetalia.* Riverbanks and lakeshore, wet woodland; up to 1400 m a.s.l., unsuitable for lawns, biennial to perennial.	Stem 0.4–1 m Root 0.3 m with short rhizomes	0.05	20000
Agrostis stolonifera (Creeping bent)	*Agropyro–Rumicion, Plantaginetalia.* Riverbanks, ditches, wet areas, moisture indicator, pioneer plant, dense sward, can stand heavy grazing, best grass for golf links in the northern, moist and cool areas; up to 1800 m a.s.l., perennial.	Stem 0.1–0.7 m stoloniferous, rooting at the nodes Root 0.3 m	0.05–0.09	11000–20000
Agrostis tenuis (Common bent, brown top)	*Festuco–Cynosuretum, Polygono–Trisetion, Arrhenatheretalia, Nardo–Callunetea, Sedo–Scleranthetea.* Grassy patches in woodland, moist grassland in mountainous regions, moors, recently cut forest areas, indicates acid and very poor soil conditions; valuable grass for mountainous terrain, does well in grass mixtures; up to 2200 m a.s.l., perennial.	Stem 0.2–0.4 m Root up to 0.5 m	0.06	18000

a.s.l. = above sea level.
*These plant associations are specific to Alpine regions.

Latin and English name	Plant association*, habitat and properties	Morphology: stem and root	1000 grain wt	Seeds/g
Grasses *continued*				
Alopecurus pratensis (Foxtail)	*Arrhenateretalia, Calthion, Filipendulo–Petasition, Molinio–Arrhenatheretea.* Riverbanks, alluvial areas, resistant to late frosts, overwatering, tolerant to long lasting snow cover, needs fertiliser and irrigation, then suitable for poor soils; up to 1800 m a.s.l., moisture indicator, does not stand up to heavy grazing, perennial, loosely tufted.	Stem 0.3–1 m Root up to 0.2 m	0.7–0.9	1100–1400
Anthoxantum odoratum (Sweet vernal grass)	*Molinio–Arrhenatheretea, Nardo–Calunetea, Cariceta curvulae, Quercion roboris.* Meadows and pastures, open woodland, poor mountain meadows together with *Festuca rubra, Agrostis tenuis*; up to 2500 m a.s.l., grows during winter, can stand irrigation, indicates poor soil, short lived, perennial.	Stem 0.3–0.5 m Root up to 0.5 m	0.6	1700
Arrhenatherum elatius (Oat grass)	*Arrhenatherion elatioris, Calamagrostidion arundinacii.* Moist and fertile meadows, at higher elevations replaced by *Trisetum sp.*, main species of fertilised meadows, avoids wet heavy and compacted soils, sensitive to moist–cool conditions; up to 1500 m a.s.l., perennial, loosely tufted.	Stem 0.5–1.5 m Roots tough, deep rooting	3.3	300
Avena sativa (Oats)	Best cover crop in humid habitats, frost tender, high water demand, desiccates the soil; up to 1600 m a.s.l., annual, at high elevations in the mountains, bi-annual to perennial.	Stem up to 0.8 m Root up to 0.3 m	33.3	30
Avenella flexuosa (Oat grass, flexible)	*Vaccinio–Piceion, Eu-Vaccinio–Piceion, Epilobion angustifolii, Luzulo–Fagion, Quercion roboris, Nardo–Callunetum.* Acid humus pioneer on poor soils, shade tolerant; up to 2200 m a.s.l., perennial.	Stem 0.3 m, slender Root up to 0.5 m, tough	1.25	1600

Species	Description	Size	Value 1	Value 2
Brachypodium pinnatum (Chalk false brome)	Meosbromion, Cirsio–Brachypodion, Festuco–Brometea, Nardetalia, Molinion dry, Erico–Pinion, Cephalanthero–Fagion. Indicates base-rich soil and deteriorating woodland conditions, enhanced by burning, weakened by fertiliser application; up to 1600m a.s.l., perennial.	Stem 0.6–1.2m Root up to 0.5m, tough	4.6	220
Bromus erectus (Upright brome)	Xerobromion, Mesobromion, Molinion, Arrhenatherion, Salvia–Trisetetum. Semi-arid grassland on limestone, in the south on gneiss and serpentine, pioneer, intolerant of fertiliser and irrigation, avoids shade and wetness, but very resistant to dry heat; up to 1400m a.s.l., perennial, tufted.	Stem up to 0.6m Root up to 0.8m, tough	4.6	220
Bromus inermis (Hungarian brome)	Cirsio–Brachypodietum, Arrhenatherion, Sisymbrion. Pioneer, very drought and cold resistant; up to 1200m a.s.l., perennial.	Stem 0.3–1.4m Rhizomatous, deep rooted	2.2–3.3	300–450
Bromus mollis (Soft brome)	Bromo–Hordeetum, Sisymbrion, Arrhenatherion. Grassland weed, indicates poor soil, suitable cover crop on dry sites; up to 1000m a.s.l., annual.	Stem 0.2–0.8m Root up to 0.2m	3	325
Cynodon dactylon (Couch grass)	Polygonion avicularis, Cynosurion. Pioneer for the quick stabilisation of sandy soil in low altitudes, starts late in the season, frost hardy, pasture grass, turns brown in winter; up to 1000m a.s.l., perennial.	Stem procumbent with long stolons Rhizomatous	0.3–0.58	1700–3500
Cynosurus cristatus (Crested dog's tail)	Cynosurion, Arrhenatherion. Fertilised permanent pasture, indicates heavy soil, frost tender, shade tolerant for pastures and meadows; up to 1500m a.s.l., perennial.	Stem 0.3–0.6m	0.5–0.58	1700–2000
Dactylis glomerata (Cocksfoot)	Arrhenatherion, Mesobromion. All round pioneer, vigorous when fertilised; up to 1900m a.s.l., perennial.	Stem 0.5–1m Tufted, root up to 0.4m	1.2	900

Latin and English name	Plant association*, habitat and properties	Morphology: stem and root	1000 grain wt	Seeds/g
Grasses continued				
Deschampsia caespitosa (Tufted hair grass)	Molinion, Filipendary–Petasition, Calthion, Fagion, Alno–Padion. Intermittent moist locations in woodland and meadows, wet places and spring horizons, marshes; up to 2800m a.s.l., perennial, vigorous, stiffly tufted.	Stem 0.3–0.8m Root up to 1m	0.25	4000
Festuca arundinacea (Tall fescue)	Potentillo–Festucetum arundinaceae, Agropyro–Rumicion, Molinion, Calthion. Indicates soil compaction and wetness, tufted, tolerant of trampling, suitable for pathways and terraces in vineyards and orchards; perennial.	Stem 0.6–1.5m Rhizomatous, deep rooted	1.62–1.9	530–620
Festuca ovina (Sheep's fescue)	Festuca–Brometea, Sedo–Scleranthetea, Molinio–Arrhenatheretea, Quercion roboris, Pinion sylvestris. Dry, poor grassland on soils derived from acid rocks, indicates degradation in forests; up to 2300m a.s.l., perennial.	Dense, short grass Stem 0.15–0.4m Root up to 0.5m	0.5	2000
Festuca pratensis (Meadow fescue)	Molinio–Arrhenatheretea, Mesobromion. Fertile meadows and pastures, prefers heavy and moist soils, winter hardy, sensitive to over-use; up to 1600m a.s.l., perennial.	Stem 0.3–1.2m Loose tufts, shallow rooted	2	500
Festuca rubra subsp. rubra (Red fescue)	Molinio–Arrhenatheretea. Montane meadows and pastures, coniferous wood and deciduous forests, drought and wetness sensitive; up to 2000m a.s.l., perennial.	Stem 0.2–0.7m Runners, root up to 0.5m	0.9–1	1000–1100
Festuca rubra subsp. commutata (Red fescue)	Nardetalia, Cynosurion, Polygono–Trisetion. On acid soils, replaced by Nardus stricta if over used; up to 2000m a.s.l., perennial.	Stem 0.1–0.6m Dense sward, root up to 0.5m	1	1000

Ground Bioengineering Systems 27

Species	Description	Stem/Root		
Festuca tenuifolia syn. *capillata* (Fine-leaved sheep's fescue)	*Nardo–Galion, Thero–Airion, Quercion robori–petraeae, Castaneta*. On acid sandy soils, indicates deterioration in forests; up to 1000m a.s.l., perennial.	Stem 0.2–0.3m Slender, shallow rooted	0.4–0.6	1700–2500
Festuca trachyphylla syn. *longifolia* (Rough-leaved fescue)	*Festuco–Sedetalia, Xerobromion, Seslerio–Festucion, Koelerion glaucae*. Origin in the low lying areas of northern Germany on sandy soils, widespread in Central Europe and England; up to 1000m a.s.l., perennial.	Stem 0.1–0.15m Very shallow rooted	0.6	1700
Holcus lanatus (Yorkshire fog)	*Arrhenateretalia, Molinion, Calthion*. Indicates acid soils low in nitrogen, frost tender, green during winter, in years of good rainfall very prolific on poor soils; up to 900m a.s.l., perennial.	Stem 0.3–1.6m Tufted, root up to 0.4m	0.3–0.4	2500–3500
Holcus mollis (Creeping soft grass)	*Quercion robori-petraeae, Nardo–Callunetea*. Arable lands and ploughed out pasture, wet soil areas, never on calcareous soil, sandy soils, troublesome weed in gardens and arable land, starts growing later than *Holcus lanatus*; up to 1500m a.s.l., perennial.	Stem 0.3–1.6m Long rhizomes, root up to 0.4m	0.25	4000
Lolium multiflorum subsp. *italicum* (Italian ryegrass)	*Bromo–Hordeetum, Sisymbrion, Arrhenatherion*. Only in Atlantic areas under moist mild climatic conditions; frost damage below –5°C, needs potash-rich soils, vigorous after cuts, unsuitable for permanent meadows; up to 1700m a.s.l., annual to bi-annual.	Stem 0.3–0.9m Root up to 0.8m	2.2	470
Lolium perenne (Perennial ryegrass)	*Lolio–Cynosuretum, Polygonion avicularis*. Tolerant to repeated cutting and trampling, pioneer of well-aerated and moist soil, needs fertiliser; up to 1000m a.s.l., but in certain locations up to 2300m, fast growing perennial.	Stem 0.3–0.7m Dense tufts, root up to 1.2m	1.4–2.2	470–720

Latin and English name	Plant association*, habitat and properties	Morphology: stem and root	1000 grain wt	Seeds/g
Grasses *continued*				
Phleum pratense (Timothy)	*Cynosurion, Arrhenatheretalia.* Pastures, tolerates cold climate, wetness and prolonged snow cover, wind; grazing increases vigour and yield; up to 2600 m a.s.l., perennial.	Stem 0.2–1 m Loose tufts, roots delicate with short runners	0.5	2000
Poa annua (Annual meadow grass)	*Plantaginetalia majoris, Polygonion avicularis, Cynosurion, Chenopodietea, Secalinitea.* Tolerates trampling and heavy fertiliser applications; up to 3000 m a.s.l., annual to perennial.	Stem 0.02–0.35 m Dense low sward, shallow roots	0.5	2000
Poa compressa (Flat-stalked meadow grass)	*Alysso–Sedion, Tussilaginetum, Cirsio–Brachypodion, Festuco–Sedetalion.* Not on soils derived from acid rocks; up to 1800 m a.s.l., green over winter, perennial.	Stem 0.2–0.4 m Loose tufts with runners, root up to 0.2 m	0.15–0.18	5500–6500
Poa nemoralis (Wood meadow grass)	*Carpinion, Fagion, Quercion pubescentis-petraeae, Prunetalia.* On heavy soils, grows early after melting snow, very shade tolerant, not to be planted in pure stands, does not form close sward; up to 2300 m a.s.l., perennial.	Stem 0.2–0.9 m Tufted, shallow rooting	0.6	1700
Poa palustris (Swamp meadow grass)	*Phragmition, Magnocaricion, Phalaridetum, Calthion, Alnion.* On riverbanks, early, resistant to late frost; up to 1500 m a.s.l., perennial.	Stem 0.3–1.2 m Shallow rooting	0.2–0.25	4000–5200
Poa pratensis (Meadow grass)	*Molinio–Arrhenatheretea, Mesobromion, Festuco–Brometea.* Important constituent of meadows and pastures, wide ecological amplitude, hardy, long living, grows early in spring, very suitable for first seedings; up to 2300 m a.s.l., perennial.	Stem 0.15–0.9 m Dense tufts, rhizomatous roots to 0.65 m	0.22–0.31	3200–4500

Poa trivialis (Rough meadow grass)	*Calthion, Filipendulo–Petasition,* moist *Arrhenathereta*. On spring horizons, fens, sensitive to dry air and soil, frost and prolonged snow cover, resistant to heavy grazing, responds well to organic fertilisers; up to 1600 m a.s.l., long life, forms dense sward.	Stem 0.3–0.9 m Surface runners, shallow rooted	0.18–0.33	3000–5500
Puccinellia distans (Sea meadow grass)	*Juncetalia Blysmo–Juncetum, Puccinellion maritimae.* Saline soils, manure heaps, cattle pens, solonetzic soil; perennial.	Stem 0.15–0.5 m Tufted, root up to 0.25 m	0.24	4200
Trisetum flavescens (Yellow oat)	*Polygono–Trisetion, Arrhenatherion.* Fertile meadows in montane and subalpine regions, frost sensitive, resistant to repeated cutting; up to 2300 m a.s.l., perennial.	Stem 0.3–0.8 m Loosely tufted, root up to 0.4 m	0.26–0.5	2000–3800

Herbs and legumes

Achillea millefolium (Yarrow, milfoil)	*Arrhenatheretalia, Nardetalia, Mesobromion.* Meadows and pastures, drought resistant, indicator of fertile soil, medicinal use; up to 1900 m a.s.l., perennial.	Stem 0.2–0.6 m Rhizomatous, root up to 4 m	0.15	6700
Chrysanthemum leucanthemum (Ox-eye, dog daisy)	*Arrhenatheretalia, Molinietalia, Mesobromion.* Pioneer on loose, well-aerated, immature soils, indicates poor fertility in meadows; up to 2200 m a.s.l., perennial.	Stem 0.3–0.6 m Root up to 0.6 m	0.3–0.38	2600–3300
Pimpinella saxifraga (Burnet saxifrage)	*Xero–Mesobrometea, Nardo–Galion, Erico–Pinion, Festuco–Brometea.* Indicates poor, dry soil, useful in sheep pasture on calcareous soil; up to 2300 m a.s.l., perennial.	Stem 0.15–0.5 m Root up to 1.3 m deep, 8–10 m long	1.5–9	110–670
Plantago lanceolata (Plantain)	*Molinio–Arrhenaheretea.* Worldwide distribution on many soil types, even in waterfowl runs, very sensitive to herbicides; up to 1800 m a.s.l., perennial.	Stem 0.05–0.5 m Root up to 0.6 m	1.65	625

30 Ground Bioengineering Techniques

Latin and English name	Plant association*, habitat and properties	Morphology: stem and root	1000 grain wt	Seeds/g
Herbs and legumes *continued*				
Sanguisorba minor (Salad burnet)	Mesobromion, Festuca–Brometea, Arrhenatherion, Erico–Pinion, Brometalia erecti. Pioneer plant, root fungus symbiosis, medicinal use, spice; up to 1200m a.s.l., perennial.	Stem 0.3–0.6m Root up to 1.5m	1.3–9	110–815
Anthyllis vulneraria (Kidney vetch)	Mesobromion, Cirsio–Brachypodion, Xerobromion, Molinion, Erico–Pinion, Arrhenatherion. Cannot take fertiliser application or irrigation, pioneer, frost and drought resistant; up to 2000m a.s.l., perennial.	Stem 0.1–0.5m Root over 1m	2.5	400
Coronilla varia (Crown vetch)	Mesobromion, Arrhenatherion, Onopordion. On dry sunny slopes; up to 900m a.s.l., perennial.	Stem 0.3–1.2m Root up to 0.9m	4	260
Lotus corniculatus (Bird's foot trefoil)	Mesobromion, Trifolion medii, Arrhenatheretalia, Molinion. Semi-dry turf, fertile meadows and pastures, prefers calcareous soils, high temperature resistance; up to 2300m a.s.l., perennial, persists for 20 years.	Stem 0.05–0.6m Tap root up to 1m	1–1.3	750–1000
Lotus uliginosus (Large bird's foot trefoil)	Calthion, Molinion, moist Arrhenatheretea, Alno–Padion. Moist meadows and pastures, spring horizons, bog plant, indicates nitrogen-rich soil; up to 1000m a.s.l., perennial.	Stem 0.3–0.9m Root more than 1m	1–1.3	750–1000
Lupinus albus (White lupin)	Up to 600m a.s.l., annual.	Stem 0.2–1m Root up to 0.75m	33.3	30
Lupinus luteus (Sweet lupin)	Up to 1400m a.s.l., annual.	Stem 0.3–1.2m	33.3	30
Lupinus polyphyllus (Garden lupin)	Sambuco–Salicion. On woodland fringes and clearings; up to 1400m a.s.l., perennial.	Stem 1–1.5m Root more than 1m	22.2	45

Species	Description	Dimensions	
Medicago falcata (Sickle medick)	*Festuco–Brometea, Geranion sanguineae–Brachypodion, Arrhenatherion*. Not cultivated as it becomes woody. *Medicago varia* (= *M. falcata* × *sativa*) traded under the name 'lucerne'; up to 1100m a.s.l., perennial.	Stem 0.2–1m Deep rooted	2	500
Medicago lupulina (Black medick)	*Mesobromion, Caucalion, Lolio–Cynosuretum, Arrhenatheretalia*. Dry meadows of good fertility, indicator of dry habitat, undemanding pioneer, prefers calcareous soil, frost resistant, can be heavily grazed; up to 1500m a.s.l., annual to perennial.	Stem 0.1–0.6m Thin tap root, up to 0.5m	1.8–2.3	435–550
Medicago sativa (Lucerne)	*Mesobromion* and dry *Arrhenatherion*. Native of Iran, today only available as hybrid, sensitive to late frosts; up to 1000m a.s.l., perennial.	Stem 0.3–1.2m Very tough tap root, 2.5–5m (up to 10m)	0.7–2.5	400–600
Melilotus albus (White melilot)	*Echio–Melilotetum*. Drought resistant, becomes woody, needs mowing; up to 1800m a.s.l., bi-annual.	Stem 0.3–1m Thick tap root, 0.7m	1.8	570
Melilotus officinalis (Common melilot)	*Echio–Melilotetum, Tussilaginetum, Caucalion*. Used medicinally; up to 1000m a.s.l., bi-annual.	Stem 0.3–1m Tap root, up to 0.75m	1.8	570
Onobrychis viciifolia (Common sainfoin)	*Mesobromion, Brometalia erecti*. Dry soil indicator in *Arrhenatheretum*; important feed plant on dry, clayey calcareous soil, sensitive to grazing pressure; up to 2000m a.s.l., lasts 4–6 years.	Stem 0.1–0.7m Root 1–4m	20–29	35–50
Phacelia tanacetifolia	Up to 1000m a.s.l., useful as green crop, annual.	Stem up to 0.7m Root up to 0.2m		
Pisum sativum (Garden pea)	Up to 1000m a.s.l., cover and green crop, annual.	Stem 0.5–2m Tapering tap root	143–500	2–7
Trifolium dubium (Lesser clover)	*Arrhenatheretea, Cynosurion, Arrhenatheretum*. Fertile meadows and pastures, needs heavy nitrogen dressings, up to 1000m a.s.l., annual to bi-annual.	Stem 0.05–0.35m Root up to 0.2m	0.5–0.55	1850–2000

Latin and English name	Plant association*, habitat and properties	Morphology: stem and root	1000 grain wt	Seeds/g
Herbs and legumes continued				
Trifolium hybridum (Alsike clover)	Arrhenatherion, Bromion racemosi, Molinion, Calthion, Agropyro–Rumicion. Pioneer on moraines, tolerates moist and cool conditions, frost resistant, tolerates prolonged snow cover, sensitive to shade and dry soil conditions; up to 2000m a.s.l., bi-annual to perennial.	Stem 0.2–0.7m Root 0.2–0.8m, much branched	0.6–0.9	1100–1600
Trifolium pratense (Red clover)	Arrhenatheretalia, Calthion, Molinion, Mesobromion, Eu–Nardion. Fertile meadows and pastures, but also moist and poor meadows, sensitive in spring to grazing pressure, important feed plant; up to 2200m a.s.l., perennial.	Stem 0.2–1.2m Root to 2m, much branched	1.5–2.3	450–670
Trifolium repens (White clover)	Cynosurion, Plantaginetalia. Heavily used turf, meadows, parks, aerodromes in humid areas, very prolific, up to 2300m a.s.l., perennial.	Stem 0.1–0.5m Root 0.7m, rooting at nodes	0.6–0.8	1250–1700
Vicia sativa (Common vetch)	Brometea, Chenopodietea, Secalinetea. Valuable cover crop; up to 1600m a.s.l., annual, many varieties.	Stem 0.3–1m, prone or climbing Root up to 0.5m	46	22
Vicia villosa (Fodder vetch)	Papaveretum argemone, Secalinion. Cover crop; up to 1700m a.s.l., frost hardy if sown early, annual to bi-annual.	Stem 0.3–0.6m, prone or climbing Root up to 0.6m	29	35
Conifers				
Larix decidua (European larch)	Constituent of coniferous forests, but also in mixed woodland, pure stands in certain localities in the Western and Southern Alps; main distribution in the alpine-continental Spruce and Arolla pine forests up to the tree line at 2100–2400m a.s.l.	Up to 35m Decidous taproot system	5.9	170

Species	Description	Size / Roots		
Picea abies (Norway spruce)	Pure and mixed stands on moist, humic, slightly acid soil derived from siliceous and calcareous rocks. *Vaccinio–Piceion, Eu–Fagion*. Optimum in the montane and subalpine region between 800–1900 m a.s.l.	Up to 30 (35)m, evergreen Dominant, widespread shallow roots, in very deep soil some vertical roots	7.7	130
Pinus sylvestris (Scots pine)	In mixed multi-species pine forests on base-rich calcareous soil (*Erico–Pinetum, Seslerio–Pinetum, Carici–Pinetum, Ononido–Pinetum*) and in acid pine forest on poor acid soil (*Calluno–Pinetum, Astragalo–Pinetum, Vaccinio–Pinetum*). Pure and mixed stands up to 1600 (1900)m a.s.l., drought resistant, frost hardy pioneer tree.	Up to 15 (20)m, evergreen Tap root	6.6	15
Pinus uncinata (Mountain pine)	Montane and subalpine forest on very shallow, stoney, calcareous soils. *Pinetum montanae*. Optimum in the Western Alps, in the northern limestone Alps of the east-Alpine region eastward to Berchtesgaden; Pyrenees, up to tree line at 2000–2400m a.s.l., locally in small patches in montane and subalpine boggy areas; hardy pioneer.	Up to 15(20)m, evergreen Tap root system	6.6	15

Broadleaved trees

Species	Description	Size / Roots		
Acer platanoides (Norway maple)	In mixed woodland from low lying areas up to 1100m a.s.l., on moderately acid soils, immature soils.	Up to 30m Deep rooted	125	8
Acer pseudoplatanus (Sycamore)	In moist, cool, mixed broadleaved tree woodland from peri-alpine hills to 1700m a.s.l., requires humus-rich soil adequately supplied with water, tolerant to long periods of gravel aggradation.	Up to 25m Deep rooted	83	12

Latin and English name	Plant association*, habitat and properties	Morphology: stem and root	1000 grain wt	Seeds/g
Broadleaved trees continued				
Alnus glutinosa (Common alder)	Forest pioneer of fens and riverbanks. *Alnion glutinosae, Alno–Padion*. On acid, moist to wet soil up to 1500m a.s.l., pioneer on immature soil, nitrogen fixing.	Up to 20m Root development depending on habitat, but always deeper than grey alder	1.25	800
Alnus incana (Grey alder)	Pioneer on alluvial immature soils, riverbanks in the montane region. Forms single species woodland on *Alno–Padion, Alnetum incanae* on alluvial flood plains, up to 1400 (1600)m a.s.l., nitrogen fixing.	Up to 20m Shallow rooted	0.68	1470
Betula pendula (Silver birch)	Pioneer of most forest types in Central Europe, particularly on sandy and poor soil, sub-dominant in open spruce–oak–beech–alder woodland; up to 1700m a.s.l., particularly suitable for mine dumps.	Up to 15(20)m Intensive shallow root system	0.14	7140
Betula pubescens (Hairy birch)	Pioneer on acid, wet peaty soil and immature soils on siliceous parent material in humid areas; up to 1800m a.s.l.	Up to 12m Intensive shallow root system	0.26	3846
Fraxinus excelsior (Ash)	May form pure stands in hardwood forests and in gorges (*Alno–Padion, Fagion*). Formerly important browse tree, therefore widespread plantings up to 1400m a.s.l., sensitive to late frost, important soil stabilising properties.	Up to 35m Extensive, deep root systems, tough roots	70	14
Fraxinus ornus (Flowering ash)	In the warmer zones of broadleaved woodlands (*Orno–Ostryon*) in the Rhine valley, lower slopes of the southern Alps and pannonian areas up to 800m a.s.l.	Up to 8m Deep rooted	65	15

Species	Description	Size/Roots		
Prunus avium (Wild cherry, gean)	Constituent of open broadleaved woodland (*Carpinion, Fagion, Ulmion*). At higher elevations only on forest margins; up to 1300m a.s.l., planted up to 1700m.	Up to 15m Deep rooted	166–200	5–6
Prunus padus (Bird cherry)	Constituent of mixed broadleaved woodland (*Alno–Padion, Fagetalia*) at higher elevations. Intolerant to shading; from low lying areas up to 1700m a.s.l., on rich and fertile soils, resistant to flooding and gravel aggradation.	Up to 15m Extensive root system with tough roots	45.5	22
Sorbus aria (Common whitebeam)	Sunny positions on base-rich soils in broadleaved and coniferous woodland in warmer areas; (*Quercion pubescentis, Fagetalia, Berberidion*); from low lying areas to 1500m a.s.l.	Up to 12m, reaches age of 200 years Deep rooted	3.75–6	167–267
Sorbus aucuparia (Rowan)	Wide-spread in almost all humid forest types; from low lying areas up to 1800 (2000)m a.s.l., tolerates light shade only, on any soil type.	Up to 15m Deep rooted, on deep soil	2.5–3	330–400
Shrubs				
Acer campestre (Field maple)	Common in broadleaved forests and spinneys (*Carpinion, Ulmion, Cephalanthero–Fagion, Acerion, Quercion pubescentis, Berberidion*); up to 800m a.s.l., tolerates only light shade.	Slow growing shrub to small tree, suitable for hedges, suckers freely Widespread tough root system	83	12
Alnus viridis (Green alder)	Forms closed formations on montane and subalpine alluvial flood plains and shelves; (*Adenostylion, Alnetum viridis*). On wet, lower slopes in snow-rich areas; from 500–1800m a.s.l.	Up to 3m Shallow rooted soil improver under prolonged snow cover, procumbent, impedes tree growth	0.55–0.62	1600–1800

36 Ground Bioengineering Techniques

Latin and English name	Plant association*, habitat and properties	Morphology: stem and root	1000 grain wt	Seeds/g
Shrubs continued				
Amelanchier ovalis (Snowy mespilus)	Occasional in sunny oak and pinewoods (*Quercion pubescentis*, *Pinetum sylvestris*, *Berberidion*), on rocky, stony slopes; up to 1800m a.s.l.	Up to 3m, showy Roots in rock fissures, widespread root system	76.9–83.3	12–13
Colutea arborscens (Bladder senna)	In sunny, warm locations of oakwoods (*Quercion pubescentis*, *Lithospermo–Quercetum*, *Berberidion*) in the south; up to 800m a.s.l., on any soil type.	Up to 3m Deep rooted	1	1000
Cornus mas (Cornelian cherry)	Sunny, open woodland and woodland fringes (*Quercion pubescentis*, *Berberidion*, *Alno–Padion*); up to 600m a.s.l., tolerates light shade, suitable hedge plant.	Up to 4m high shrub or small tree to 6m Widespread tough root system	166.6	6
Cornus sanguinea (Dogwood)	In sunny, open broadleaved mixed woodland, wood verges (*Prunetalia*, *Alno–Padion*, *Carpinion*, *Quercion pubescentis*, *Cephalanthero–Fagion*); up to 1000m a.s.l., any soil type, light shade only.	Up to 3.5 (5) m, spreading suckers readily Widespread tough root system	30.3	33
Corylus avellana (Hazel)	Secondary formations in the temperate broadleaved mixed woodland zone (*Carpinion*, *Alno–Padion*, *Prunetalia*, *Querco–Fagetea*) in sunny, sub-mediterranean/sub-atlantic locations; up to 1400m a.s.l., on many soil types, tolerates moderate shade.	Shrub or small tree to 5 (6.5)m, spreading habit Extensive tough root system	1000	1
Crataegus monogyna (Common hawthorn) *Crataegus oxyacantha* Midland thorn)	Constituent of sub-atlantic and sub-mediterranean broadleaved and coniferous woodland (*Quercion*, *Carpinion*, *Pinion*, *Ulmion*, *Fagion*); up to 1000m a.s.l., may be clipped annually, suitable for hedges.	Up to 6 (10)m, occasional as tree when trained, thorny Deep rooted Can reach 100 years of age	8.3	125

Species	Habitat/Description	Size and root system	
Cytisus scoparius (Broom)	On acid, non-calcareous soil on sunny slopes in areas of mild winters, (*Calluno–Sarothamnion, Sambuco–Salicion, Carpinion, Quercetalia roboris*); from low lying areas to 1100 m a.s.l., pioneer on immature soil, nitrogen fixing.	Up to 2 m deep Extensive root system	8 120
Euonymus europea (Spindle tree)	Thickets and woodland fringes, open woodland (*Pinion, Alno–Padion, Carpinion, Fagion*); from low lying areas to 1100 m a.s.l., on calcareous soil and sunny positions, indicates loamy soil, semi-shade.	Up to 4 m shrub or small tree Extensive root system	23–25 40–43
Frangula alnus (Alder buckthorn)	Common in bogs and damp woods, riverbanks, alluvial areas, open oak and pinewoods (*Alnion, Molinion, Alno–Padion, Quercion roboris, Luzulo–Fagion, Calluno–Genistion*); up to 1000 (1300) m a.s.l., tolerates flooding and oxygen-poor soil due to compaction.	2–7 m, loosely erect, occasionally small tree Shallow rooted with root bulbils	30 33
Genista germanica (German greenweed)	On acid soil of sunny slopes in the peri-alpine area (*Calluno–Genistion, Quercion roboris*); up to 750 m a.s.l., indicates progressively acid soil conditions.	Up to 1.5 m Spreading root system	3.2 312
Genista tinctoria (Dyer's greenweed)	Common on poor soils on sunny slopes in the peri-alpine areas, infertile meadows, oak forests (*Molinion, Calluno–Nardetum, Quercion roboris*); lowlands up to 750 m a.s.l., indicates alternate wet–dry conditions.	Up to 1 m Root up to 1 m	3.3 300
Hippophae rhamnoides (Sea buckthorn)	Closed formations, pioneer vegetation on gravelly alluvia in the peri-alpine area, in dry pine forests on sandy soil and sand dunes of the North and Baltic Seas (*Berberidion, Erico–Pinion, Alnetum incanae*); up to 1000 m a.s.l., forms root nodules, intolerant to shade.	Up to 3 (5) m, either small trees or thicket forming, suckers freely	7.5 133

Latin and English name	Plant association*, habitat and properties	Morphology: stem and root	1000 grain wt	Seeds/g
Shrubs *continued*				
Laburnum anagyoroides (Golden rain)	Constituent of sub-mediterranean oak woods and pine forests (*Quercion pubescentis, Lithospermo–Quercion*); up to 1000m a.s.l., soil improver, can be vegetatively propagated.	Large shrub to 7 m, ascending branches Strong widespread root system	14–32	30–70
Laburnum alpinum (Alpine laburnum)	Constituent of montane and subalpine beech–fir tree woodland (*Fagion*) in humid areas, in the Alps replaces the green alder, vegatively propagated.	Up to 5m, ascending branches	20–40	25–50
Ligustrum vulgare (Privet)	In sunny broadleaved and coniferous woodland on neutral- to base-rich soil (*Berberidion, Quercion pubescentis, Erico–Pinion, Alno–Padion, Carpinion*); pioneer plant up to 1000m a.s.l., vegetatively propagated, can be clipped (hedges).	Up to 2(3)m Extensive roots, runners	40	50
Lonicera xylosteum (Fly honeysuckle)	Common shrub in broadleaved, coniferous mixed woodland (*Fagion, Quercion pubescentis, Prunion*); from low lying areas to 1100(1600)m a.s.l., tolerates moderate salinity and shade.	Up to 2 m loosely spreading Shallow rooted	10	100
Prunus mahaleb (St. Lucia cherry)	Sunny slopes and oak–pine forest in warmer areas (*Quercion pubescentis, Lithospermo–Quercetum, Berberidion*); up to 800m a.s.l., semi-shade.	Up to 4m loosely spreading, rarely a small tree Deep rooted	90–100	10–11
Prunus spinosa (Sloe, blackthorn)	In sunny positions, woodland fringes and open woodlands, (*Prunetalia*); up to 1000m a.s.l., pioneer on immature soils, intricately branched.	2–3m Creeping roots	1000	1

Ground Bioengineering Systems 39

Species	Habitat/Notes	Growth characteristics		
Rhamnus cathartica (Buckthorn)	Singly in wood verges and sunny spinneys, (*Prunetalia, Quercetum pubescentis*), mostly on calcareous soil; lowlands up to 1300 m a.s.l.	2–3 m, much branched, thorny shrub or small tree up to 6 m, slow growing. Extensive root system	14.3	70
Rosa canina (Dog rose)	Common in sunny positions, spinneys (*Prunetalia*); up to 1300 m a.s.l., on many soil types.	Up to 3 m, loosely spreading. Deep rooted	2.8–3.3	30–35
Rosa rubiginosa (Sweet-briar)	Cosmopolitan in sunny positions, spinneys and woodland in warmer areas, prefers calcareous soils (*Berberidion, Prunetalia*); up to 1200 m a.s.l., indicates loamy soil.	Up to 3 m, loosely spreading, thorny. Deep rooted	10	100
Rubus fruticosus (Blackberry)	Forest floor pioneer in humid areas with mild winters; lowlands up to 1600 m a.s.l.	Up to 1 m, arching branches, fast spreading. Rootshoots	2	500
Sambucus nigra (Elder)	Common in damp woodland, wasteland, spinneys, prefers damp and fertile soils (*Alno-Padion, Fagetalia, Sambuco–Salicion, Brometalia, Robinia* woodland); up to 1200 m a.s.l., indicates nitrogen-rich soil.	Up to 5 m, much branched, wide crown, shrub or small tree. Shallow rooted	2.5	400
Sambucus racemosa (Alpine elder)	Often in damp and shady woods, cleared woodland, spinneys, in the montane region (*Fagion, Berberidion*); lowlands up to 1800 m a.s.l., usually on non-calcareous soil, nitrate indicator.	Up to 3 m, tall spreading bush. Shallow rooted, new shoots produced from roots	7	143
Viburnum lantana (Wayfaring tree)	Scattered in open pine and oak woods, spinneys on calcareous soils (*Ligustro–Prunetum, Berberidion, Quercion pubescentis, Erico-Pinion*); up to 1400 m a.s.l., sprouting branches, tolerates severe cutting, hedge plant.	Up to 3 (4) m, spreading. Extensive root system	43.4–45.4	22–23
Viburnum opulus (Guelder rose)	Common on alluvial shelves, bushes and broadleaved woods (*Prunetalia, Alno-Padion*), hydromorphic deep alluvial soil; up to 1000 m a.s.l., cut branches shoot readily, indicates moving water table.	Up to 5 m, fast growing, large bush or small tree. Extensive shallow root system	40	25

Table 2.3 Vegetatively propagated trees and woody plants suitable for vegetative methods.

Latin and English name	Size	Habitat	Vegetative propagation % take
Tree			
Populus nigra (Black poplar)	Up to 30 m	Softwood alluvials, hydromorphic but well aerated sandy silty soils, periodically flooded; up to 1000 m a.s.l., in the southern Alpine region to 1400 m a.s.l.	70–100, but only end cuttings with terminal bud, preferably from suckers
Salix alba (White willow)	Up to 20 m	Softwood alluvials, lowland/lower montane regions, neutral fertile and calcareous alluvial sandy loams and loamy sands subject to periodic flooding, tolerates silt aggradation; up to 900 m a.s.l., in the Southern Alps to 1300 m a.s.l.	Approx. 70
Salix alba subsp. *vitellina* (Golden willow)	Up to 20 m	Cultivated willow with yellow or reddish-orange young twigs; only used as ornamental on site.	Approx. 70
Salix daphnoides (Violet willow)	Up to 15 m	Softwood alluvials of mountain streams, particularly in the montane zone of the limestone Alps; on loamy, gravelly–sandy, neutral- to base-rich alluvial soils; up to 1300 m a.s.l., neutral central regions to 1850 m a.s.l.	100
Salix fragilis (Crack willow)	10–25 m	Permanently wet soils, poor base status, moving water table in areas with cool summers; alluvials in higher lying areas, can tolerate stagnant water table and gley conditions; up to 600 (1100) m a.s.l.	70–100
Salix pentandra (Bay-leaved willow)	Up to 12 m	Coppices and alluvial woodland on wet, slightly acid alluvial soil of restricted permeability, mainly in lowlands and inner-alpine valleys; up to 1800 m a.s.l.	100
Salix rubens	Up to 25 m	Softwood alluvials.	70–100

a.s.l. = above sea level.
*These plant associations are specific to Alpine regions.

Shrubs

Laburnum alpinum (Alpine laburnum)	Up to 5 m	In the Southern Alps in warm, moist woodland and bushes, on stony, rocky sites; 500–1900m a.s.l.	70–100
Laburnum anagyroides (Golden rain)	3–8 m	Oak and pine woods in the Alps, mainly southern Alps on fertile and calcareous, humous, sandy to stony loamy soils in not too dry locations, mild winters.	70
Ligustrum vulgare (Privet)	Up to 3 m	In warm broadleaved woods and pine woods, shrub associations; from lowland to 1000m a.s.l.	70–100
Salix appendiculata (Goat willow)	Up to 4 (6) m	From peri-alpine areas to the tree-line in humid locations on base-rich, neutral to slightly acid soil or calcareous rubble; (500) 1200m to 2000 (2100)m a.s.l.	50–70, but strict adherence to winter dormancy
Salix aurita (Round-eared willow)	Up to 2.5 m	From lowland to the montane regions, rare in areas of pronounced continental climate, marshes, on acid peaty gley soils; up to 1600m a.s.l.	50–70
Salix cinerea (Grey sallow)	2–3 m	From coastal marshes to mountainous regions, in the warmer areas in drying marshes, fens; alder coppices, on fertile acid sandy soils to clay soils, tolerates waterlogged conditions (gley soils); up to 800m a.s.l.	50
Salix eleagnos (Hoary willow)	Up to 6 m, rarely 15 m	Characteristic constituent of *Hippophae rhamnoides–Alnus incana* woodland–bush, pioneer in the Erica-pine forest and grey alder alluvial woodland: mainly on alluvial shelves of alpine river valleys; calcareous rubble just above the ground water table, but periodically dry, sandy–stony slides on steep slopes.	50–70, but strict adherence to winter dormancy
Salix foetida	1.5 m	Subalpine region of the central western Alps, in willow and green alder woods, riverbanks and wet slopes, on acid, fertile, often marshy soil, by preference on moraines of siliceous rocks; 1700–2000m a.s.l.	50–70, slow growing

Latin and English name	Size	Habitat	Vegetative propagation % take
Shrubs continued			
Salix glabra	1.5 m	Limestone mountains of the eastern Alps, on rubble, stony slopes and gullies, only on calcareous and dolomitic rocks; 1400–2000 m a.s.l.	70–100, slow growing
Salix glaucosericea	Up to 1.5 m	Scattered in the subalpine region of the central Alps in green alder woods and scrub on wet siliceous moraines and alluvial shelves; 1700–2000 m a.s.l.	70
Salix hastata (Mountain willow)	Up to 3 m, rigid branches	In the high montane and subalpine regions of the Alps in willow–green alder scrub–bush areas in moist and shaded locations, neutral to weakly acid, fertile soils on various parent rocks, tolerates prolonged snow cover; 1600–2100 (2400) m a.s.l.	70–100
Salix hegetschweileri	Up to 4.5 m	Subalpine and montane region of the central Alps on wet, usually weakly calcareous but fertile soils subject to percolating water, riverbanks and wet lower slopes of bayleaf willow bush areas; 1600–2000 m a.s.l.	70–100
Salix helvetica (Swiss willow)	Up to 1.5 m	Subalpine region of the central alps, on the shaded scree slopes with dwarf shrubs and green alder scrub, non calcareous, wet rubble, skeletal soils, tolerates prolonged and deep snow cover; 1700–2600 m a.s.l.	50–70, slow growing
Salix mielichhoferi	Up to 4 m	In the central part of the eastern alps on wet slopes and riverbanks in the montane and subalpine zone on fertile skeletal soils; 1300–2200 m a.s.l.	70–100
Salix nigricans (Dark-leaved willow)	Up to 8 m bush or small tree	On moist, neutral to weakly acid clayey, gravelly or sandy soil, *particularly* in the cool-humid limestone areas; up to 1600 m a.s.l.	100

Salix nigricans subsp. *alpicola*		The sub-species *alpicola* forms closed stands in the high-montane and subalpine regions of the central Alps, tolerates waterlogged conditions and shade.	
Salix purpurea (Purple osier)	Up to 6 m	Softwood alluvial bush, often in the pioneer stage, periodically flooded, usually calcareous alluvial, silty, sandy, gravelly soils; to 1600 m (2300) m a.s.l.	100 most suitable willow for vegetative protection measures
Salix triandra (Almond-leaved willow)	2–4 m	Softwood alluvial woodland, particularly in the pioneer stage, periodically flooded, wet, often calcareous, silty, sandy or gravelly soils; from lowlands to 1500 m a.s.l., not very tolerant to shade, mainly in peri-alpine river valleys.	70–100
Salix viminalis (Common osier)	Up to 5 m	Peri-alpine river valleys, often planted and therefore in the alpine river valleys; up to 1400 m a.s.l., on intermittently wet, base-rich and fertile silty–loamy sands.	70–100
Salix waldsteiniana	Up to 1.5 m	In subalpine zones of the eastern Alps on moist, neutral to weakly acid, base-rich, loamy, skeletal soils, tolerates prolonged and deep snow cover; 1400–2200 m a.s.l.	70–100 slow growing

There is evidence that under certain, not sufficiently researched circumstances, *Salix caprea*, *Alnus incana* and *Alnus glutinosa* were used with varying success. In general, this cannot be recommended at the present time.
The bracketed numbers under 'Size' refer to higher elevations associated with the sub-Alpine climatic regime.

Table 2.4 The most important rooted woody plants suitable for vegetative methods of construction.

Latin and English name	Elevation a.s.l.	Pioneer plant	Resistance to rockfall
Conifers			
Larix decidua (**European larch**)	foothills to subalpine, up to 2300 (2400)m	••	••
Pinus sylvestris (**Scots pine**)	foothills to subalpine, up to 1600 (1900)m	••	
Pinus uncinata (Mountain pine)	submontane to subalpine, up to 2300 (2400)m	•	
Broadleaved trees			
*Acer campestre** (Field maple)	foothills to submontane, up to 1000m		•••••
Acer platanoides (Norway maple)	foothills to submontane, up to 1100m		•••••
Acer pseudoplatanus (Sycamore)	submontane to subalpine, up to 1700m		•••••
Alnus glutinosa (**Common alder**)	foothills to submontane, up to 1050m	•••	•••••
*Alnus incana** (**Grey alder**)	submontane to montane, 500–1600m	•••	•••••
Betula pendula (Silver birch)	montane to subalpine, 1100–1800m		
Betula pubescens (Hairy birch)	montane to subalpine, 1100–2100m		
Carpinus betulus (**Hornbeam**)	foothills to submontane, up to 1000m		•••••
Castanea sativa (Sweet chestnut)	foothills to submontane, up to 700 (1000)m		•••••
Fraxinus excelsior (**Ash**)	foothills to montane, up to 1400m	•••	•••••
Populus alba (**White poplar, abele**)	foothills to submontane, up to 800m		
Populus nigra (**Black poplar**)	foothills to submontane, up to 800m		
Populus tremula (**Aspen**)	submontane to montane, up to 1400m	•••	
Prunus avium (Wild cherry, gean)	foothills to submontane, up to 1300 (1700)m		
Prunus padus (Bird cherry)	submontane to subalpine, up to 1700m		
Quercus petraea (Durmast, sessile oak)	foothills to submontane, up to 1000m		
Quercus robur (Common oak)	foothills to montane, up to 1200m	•••	
Salix alba (**White willow**)	foothills to submontane, up to 900 (1300)m	•	•••••••
*Salix caprea** (**Sallow**)	foothills to subalpine, up to 1700m	•	•••••••
Salix daphnoides (**Violet willow**)	submontane to subalpine, up to 1300 (1850)m	•	•••••••
Salix fragilis (**Crack willow**)	foothills to submontane, up to 1100m	•	•••••••
Salix pentandra (**Bay-leaved willow**)	foothills to subalpine, up to 1800m	•	•••••••
Sorbus aria (Common whitebeam)	foothills to montane, up to 1500m	•	•

Ground Bioengineering Systems   45

Species	Habitat range
***Sorbus aucuparia* (Rowan)**	foothills to subalpine, up to 1800 (2000)m
Tilia cordata (Small-leaved lime)	foothills to montane, up to 1450m
Ulmus glabra (Wych elm)	foothills to montane, up to 1400m
***Ulmus minor* (Smooth-leaved elm)**	foothills to submontane, up to 600m

Shrubs

Species	Habitat range
***Alnus incana* (Grey alder)**	submontane to montane, 500–1600m
***Alnus viridis* (Green alder)**	montane to subalpine, up to 1800 (2000)m
Berberis vulgaris (Common barberry)	foothills to submontane, up to 1800m
Clematis vitalba (Traveller's joy)	foothills to submontane, up to 1000m
***Cornus mas* (Cornelian cherry)**	foothills to submontane, up to 600m
***Cornus sanguinea* (Dogwood)**	foothills to submontane, up to 1000m
Corylus avellana (Hazel)	foothills to montane, up to 1400m
***Crataegus monogyna* (Common hawthorn)**	foothills to submontane, up to 1000m
Evonymus europaeus (Spindle tree)	foothills to submontane, up to 1100m
***Hippophae rhamnoides* (Sea buckthorn)**	foothills to submontane, up to 1000m
Laburnum alpinum (Alpine laburnum)	submontane to subalpine, up to 1900m
Laburnum anagyroides (Golden rain)	foothills to submontane (subalpine), up to 1000 (2000)m
***Ligustrum vulgare* (Privet)**	foothills to submontane, up to 1000m
***Lonicera xylosteum* (Fly honeysuckle)**	foothills to montane, up to 1100 (1600)
Pinus mugo (Dwarf mountain pine)	montane to subalpine, up to 1400 (1000) to 2300m
Prunus spinosa (Sloe)	foothills to submontane, up to 1000m
Rhamnus catharticus (Buckthorn)	foothills to montane, up to 1400m
Ribes alpinum (Mountain currant)	submontane to subalpine, up to 1900m
Ribes petraeum (Flowering currant)	submontane to subalpine, up to 1900m
Rosa canina (Dog rose)	foothills to montane, up to 1350m
Rosa rubiginosa (Sweet-briar)	foothills to montane, up to 1350m
Salix appendiculata	montane to subalpine, up to 2000 (2100)m
***Salix aurita* (Round-eared willow)**	foothills to montane, up to 1600m
***Salix caprea* * (Sallow)**	foothills to subalpine, up to 1700m
***Salix cinerea* (Grey sallow)**	foothills to submontane, up to 800m

Latin and English name	Elevation a.s.l.	Pioneer plant	Resistance to rockfall
Shrubs *continued*			
***Salix eleagnos* (Hoary willow)**	submontane to montane (subalpine), up to 1400 (1850)m	•	•
Salix glabra	montane to subalpine, up to 2000m	•	•
Salix hastata	montane to subalpine, up to 2100m	•	•
Salix hegetschweileri	montane to subalpine, up to 2000m	•	•
***Salix nigricans* (Dark-leaved sallow)**	foothills to subalpine, up to 1600 (2400)m	•	•
***Salix purpurea* (Purple osier)**	foothills to subalpine, up to 2300m	•	•
***Salix repens* (Creeping willow)**	foothills to submontane, up to 1000m	•	
***Salix triandra* (Almond-leaved willow)**	foothills to montane, up to 1500m	•	
***Salix viminalis* (Common osier)**	foothills to montane, up to 1400m		
Sambucus nigra (Elder)	foothills to montane, up to 1500m		
***Sambucus racemosa* (Alpine elder)**	foothills to subalpine, up to 1800m		
***Viburnum lantana* (Wayfaring tree)**	foothills to montane, up to 1400m		
***Viburnum opulus* (Guelder rose)**	foothills to submontane, up to 1000m		
Exotics for special use			
Ailanthus altissima (Tree of heaven)	foothills, up to 500m		
Buddleia alternifolia (Buddleia)	foothills to montane, up to 800m		
Caragana arborescens (Caragana)	foothills to montane, up to 1000m		
Elaeagnus angustifolia (Elaeagnus)	foothills, up to 600m		
Forsythia intermedia (Forsythia)	foothills to montane, up to 1500m		
Lycium barbarum (Duke of Argyll's Tea-tree)	foothills to montane, up to 1200m		
Rhus typhina, *Rhus laciniata* (Sumach)	foothills to montane, up to 1000m		
Robinia pseudacacia (Robinia)	foothills to montane, up to 900m		
Rosa rugosa (Ramanas rose)	foothills to montane, up to 1000m		
Symphoricarpus racemosus (Snowberry)	foothills to montane, up to 1200m		

a.s.l. = above sea level
Bold type indicates pronounced formation of adventitious buds and resistance to secondary covering by rubble and/or soil.
An asterisk indicates species that form, depending upon environment, bushes or trees.
The bracketed numbers under 'Elevation' refer to higher elevations associated with the sub-Alpine climatic regime.
Elevations: lowland to 500m a.s.l.; montane 1100–1600m a.s.l.; submontane 500–1100m; subalpine 1600–2300m a.s.l.

Chapter 3
Ground Bioengineering Techniques for the Protection and Stabilisation of Earthworks

The classification system introduced by Schiechtl (1973) divides ground bioengineering techniques into four groups:

(1) Soil protection techniques
(2) Ground stabilising techniques
(3) Combined construction techniques
(4) Supplementary construction techniques.

Each of these groups and their corresponding construction types have definite functions and special areas of application (Tables 3.1 and 3.2).

Table 3.1 Nature and functions of ground bioengineering techniques.

Soil protection techniques rapidly protect the soil, by means of their covering action, from surface erosion and degradation. They improve water capacity and promote biological soil activity. Straw cover provides protection to soil from precipitation even before the vegetative cover (grasses, herbs, woody plants) has established itself.

Ground stabilising techniques are designed to reduce or eliminate mechanical disturbing forces. They stabilise and secure slopes liable to slides by means of root penetration, decreased pore pressure through transpiration and improved drainage. In principle, they consist of linear or single point systems of shrubs and trees, or their live cuttings, respectively. Ground stabilising techniques are generally supplemented by soil protection works to guard against erosion.

Combined construction techniques shore up and secure unstable slopes and embankments by combining the use of live plants with inert materials (stone, concrete, wood, steel, plastics). This increases the effectiveness and life expectancy of the measures employed.

Supplementary construction techniques comprise seeding and plantings in the widest sense of the word; they serve to secure the transition from the construction stage to the completed project.

Table 3.2 Applicability of ground bioengineering techniques.

	Earthworks	Riverworks	Landscaping works
Soil protection	● ●	● ●	● ●
Ground stabilisation	● ●	○ ○	○
Combined construction	● ●	● ●	●
Supplementary construction	○	●	● ●

Key: ● ● very high;   ● high;   ○ significant;
      ● moderate;   ○ ○ low;   ○ very low

3.1 Soil protection techniques (Figs 3.1 and 3.4; Plates 1–12)

Soil protection techniques are aimed in the first instance at providing easy and effective protection for the soil surface; deep soil penetration is of secondary importance. The presence of large numbers of plants, seeds or plant parts per unit surface area protects the soil from the deleterious effects of mechanical forces (raindrop impact, hail, erosion caused by water, wind or frost). Moreover, vegetation covers improve soil climate and water capacity which in turn lead to better growing conditions for succeeding plant life. They are of particular importance in areas where immediate and effective protection of the soil surface is essential (Tables 3.2 above and Table 3.3 near the end of the chapter).

3.1.1 Turfing (Fig. 3.1; Plates 1–6)

Materials
Turf can be obtained from the construction site. It should be cut as thick as possible and lifted complete with the rooted topsoil. 'Rolled turf' is available from some specialist nurseries in rolls, if necessary reinforced with wire mesh, and consists usually of a single grass variety and is cut rather thinly. It is therefore only suitable for application in low stress locations on well-prepared, moderately steep slopes with a fine seed bed. Turves that are cut by hand are rarely larger than 400 × 400 mm. The size of turves obtained by mechanical means depends upon the type of machinery used and the characteristics of the land surface; they rarely exceed 0.5 sq m. If there should be a delay between the cutting and application of the turf sections, they are to be stored in clamps not more than 1 m wide and 0.6 m high, to prevent desiccation and decomposition. Heavy infestations with mice during winter may destroy large parts of

individual clamps. During summer, the storage period should not exceed four weeks. For best results, particularly valuable or sensitive turf sections (for use in high altitude locations or nature reserves) are transported and stored on wooden pallets. Rolled turf is produced by seeding directly into some artificial growth medium which is spread on sheets and kept in shaped seed beds, or alternatively in grass nurseries, where deep soils, free of gravel and stones, are preferred for the propagation of the turf. The turf sheets can be lifted from their beds, the nursery turf is peeled by the use of suitable machinery and cut into strips usually 0.3–0.4 m wide and 1.5–2 m long, which are gathered into rolls. The thickness of the peeled strips varies between 25–40 mm.

The rolls are transported on pallets, not more than six on top of each other. Transport and storage must not exceed four days during which time the rolled turf must be protected against drying out and internal heating. Depending on the substratum, water content and thickness, rolled turf weighs approximately 25–30 kg per sq m. To calculate the required amount of turf, a shrinkage of 5% on drying must be taken into consideration. The planned production of rolled turf for a specific construction site permits the use of specially blended seed mixes, adapted to the site.

Implementation

Natural turf (Plates 1, 4 and 5)   Individual turves or rolled turf may be used to protect slope surfaces by laying them continuously and leaving no gaps. The placing of parallel strips or chequered patterns is not recommended, because the effect is still visible after many years, being obvious as something artificial in its natural surroundings. If there is only a limited amount of turf available, it is for obvious reasons best placed on the most unstable or exposed sections of the slope. On very steep slopes, each fourth or fifth turf is pegged to the ground, using 300–500 mm long stakes or reinforcement steel. The stakes must not protrude above the soil surface. The use of live stakes in dense turf is not recommended because the cuttings will not take successfully in a close sward. Fixing the turf by means of wire netting or similar plastic mesh is considered necessary on slopes which have to cope with run-off.

Rolled turf (Plates 2 and 3)   The placing of rolled turf depends upon its dimensions. Longish sections are placed vertically down the slope. On steep slopes, fixing the sheets or strips by pegging is equally important as in the case of natural turf; the distances between the pegs or stakes may, however, be greater. Rolled turf in small pieces is placed like natural turf.

After placing, the surface of the turf must be damped down or rolled. If the rolled turf has been produced by using specially adapted planting material, it can be directly placed on the slope without the need to spread a layer of topsoil first.

Grassed channels (Fig. 3.1; Plate 6)   Grassed channels serve for the safe removal of excess surface water. Flat channels, about 0.5 m deep and up to several metres wide are lined with natural turf or rolled turf, which are pegged to the ground. If large volume flows carrying base load are expected, it is of advantage to cover the turf with wire mesh; the edges of the mesh should be pulled into the soil and firmly anchored with pegs. A combination of several methods of establishing vegetative ground cover is of advantage: the edges of the channel may be reinforced by fascines or buried wattle fences.

The alignment of the channels should preferably be straight downslope. Those running obliquely across the slope have the advantage of slowing down the flow rate, but there is always the possibility that water may infiltrate into the subsoil, which may cause damage. Moreover, the construction of channels running across the slope requires a considerably higher work input. Run-off originating in the surrounding high ground should be diverted into grassed storm drains or channels that skirt the protected slope.

Turf walls   Turf walls are constructed by piling grass sods or turves to a height of up to 0.5 m. This method was fairly often used in the past, but the wall had only a chance of some permanence if it was reinforced with steel pegs, or wire mesh, etc. The practical application of turf walls is strictly limited.

3.1.2 Grass seeding (Plates 7–11)

Tremendous progress has been made during the past few years in the development of seeding methods. The application of one or other of the methods described below makes it possible to biologically activate sterile subsoils by establishing in a short space of time viable plant communities of herbs and grasses. The technical possibilities derived therefrom culminate in the simultaneous seeding of grasses and woody plants.

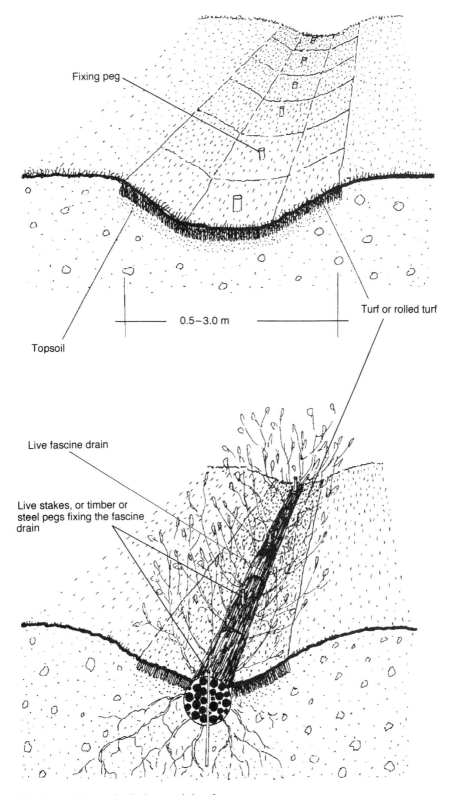

Fig. 3.1 Grassed channels. *Below:* with live fascine.

3.1.2.1 Hayseed seeding

Materials

Approximately 0.5–2 kg of hayseed per sq m are required. The hayseed is obtained from sweepings of haylofts and barns. If hayseed is to be used on its own, it is of advantage to leave a certain proportion of stems and trash as part of the mix. If the hayseed is to be used for mulch seeding, it should be screened to ensure a high amount of actual seed.

Construction

Hayseed and trash are spread to a depth of about 200–500 mm on the soil surface. To prevent the seeds from being blown away, the soil surface should be wet, or the hayseed should be moistened before use. Better results are achieved if modern seeding methods are employed. To the positive features of mulch seeding is added the advantage of using locally produced hayseed adapted to local site conditions. In mountainous terrain at high elevations, hayseed seeding will maintain its importance as long as sufficient quantities of hayseed are available.

Timing

Throughout the active growing period; for best results, during its early part.

Effectiveness

The spreading of hayseed has the same effect as other mulch seedings, provided sufficient thickness of seed material – seed plus trash – is applied; it provides soil cover for immediate protection, and has a beneficial influence on the micro-climate.

Advantages

Hayseed contains seed which is not available from nurseries and other sources; this is of particular advantage at very high elevation sites.

Disadvantages

Hayseed is only available in limited quantities, and only in places where the regular cutting of natural meadows is still practised.

Costs

The seeding costs are relatively cheap at the equivalent of 0.3–0.5 work hours per sq m, depending on procurement and transport. Hayseed seeding combined with other seeding methods adds another 10–30% to the costs.

Areas of use
The use of hayseed on its own is only feasible in Alpine regions above the timber line and at sites where commercially available seed would be inappropriate (Alpine and subalpine region). Its combination with mulch seeding or hydroseeding is particularly recommended, where the establishment of multi-species associations is the objective (permanent grassland in mountainous terrain or in very dry habitats).

3.1.2.2 Standard seeding

Materials
Short-lived plants for preparing sites for subsequent afforestation. Legumes on their own or in admixture with annual cover crop seeds which are suitable for use on immature soil or in subsoil. For permanent grass swards, only seed mixtures that are suitable for the local habitat are recommended.

Implementation
The seed is spread on the ground and lightly incorporated into the topsoil. The distance from the spreader to the soil surface on any slope should be kept as short as possible to avoid separation of the seed caused by gravity. Heavy and round seeds will roll downslope, collecting on more level ground, leaving bare patches on the steeper section. Very small and light seeds are best mixed with sand or dry clay before application. Seed incorporation on the slope is best achieved by hand raking. The use of machinery on large areas is more rational, but only feasible on flat slopes or level ground. Numerous seeders for this purpose are on the market. On flat ground that is not suitable for vehicular traffic, as for example on pipelines, on very light sandy soil, or wet marshy ground, or cleared lines in forests, the use of helicopters for seeding has proved successful. On subsoil, this method leaves seedlings without protection and is therefore less satisfactory. For this reason, standard seeding is nowadays only carried out directly onto topsoil.

Timing
During the active growing period; best at its beginning.

Effectiveness
No immediate effect. Only after termination, gradual but rapidly increasing effect through soil binding and surface cover. The use of inoculated legume seed not only provides the soil with additional

nitrogen and organic material, but also brings about rapid soil improvement after root penetration.

Advantages
A simple, rapid and cheap method.

Disadvantages
Presence of topsoil essential (in extreme Alpine conditions).

Costs
0.01–0.04 working hours per sq m; the lower figure is valid in flat areas using machinery, the higher for steeply sloping ground using hand labour. The cost of the seed has to be taken into consideration.

Areas of use
Primarily for providing temporary cover on dumps and tips that need protection against wind and water erosion, and to protect exposed topsoil from extreme desiccation. Otherwise only on level ground or gentle slopes to establish a green crop prior to afforestation.

3.1.2.3 Hydroseeding (Plate 7)

Materials
1–30 l of mix per sq m; the mix consists of seed, fertiliser, soil improvers, binding agent and water. The necessary quantities depend upon the site conditions. An assessment of the site prior to large area plantings is absolutely essential. For guidelines see Schiechtl (1973; 1980).

Implementation
Seed, fertiliser, soil improver, binding agent and water are blended in a mixer to the consistency of a thin paste. A solids pump provides the necessary pressure to spray the mix onto the soil surface. Constant agitation of the mix during operation prevents separation or settling out of components. The aim is the application of a layer approximately 5–20 mm thick which needs to be increased if applied to rough and stony surfaces. In these situations, the mix is applied in several passes, after the first application has had a chance to form a cohesive layer.

Timing
Hydroseeding should only be carried out during the rainy season on shaded positions during high atmospheric humidity. For hydroseeding with subsequent mulching, see the Section 3.1.2.5.

Effectiveness
The admixture of solids and fertiliser creates a good seed bed. The applied layer of mix forms an earthy or soil-like crust, which offers only limited resistance to mechanical forces, hydration and frost.

Advantages
Enables seeding of rocky, stony or very steep, inaccessible slopes to achieve vegetation cover quickly.

Disadvantages
Good access to the site is essential to bring the hydroseeding machine into range. Average range with extension hoses is about 150 m; without hoses it is 40 m maximum, but for most machines it is only about 25 m.

Costs
0.05–0.10 work hours per sq m, depending upon the amount of seed mix used.

Areas of use
For particularly steep, rocky and stony slopes, if accessible for the machinery required.

3.1.2.4 Dry seeding

In contrast with the hydroseeding method which uses water to apply the seed, dry seeding relies on hand application or seed blowers. In more industrialised countries, dry seeding is accomplished by the use of compressed air; alternatively seed and additives in powder or granular form are spread by helicopter. In developing countries, where cheap labour is more plentiful, application by hand is more appropriate.

3.1.2.5 Mulch seeding (Plates 8–11)

Mechanical mulch seeding

Materials
Seed, fertiliser, mulch, e.g. straw, hay, cellulose or other fibres.

Implementation
After seeding (wet or dry) a layer of chopped straw or similar material is blown onto the seeded surface, using a mechanical blower. The straw is

chopped in the mulch blower and ejected through the nozzle in one operation. On its way between chopper and nozzle, the straw is usually sprayed with an unstable bitumen emulsion to ensure rapid cohesion of the protective layer. Maximum reach is given at 25 m, increasing by the use of an extension pipe to 35 m. Greater distances are not possible, and if there is any wind, these distances will not be possible.

Timing
During the growing period.

Effectiveness
After hitting the soil surface, the straw particles stick together when the unstable bitumen emulsion sets. This provides an excellent protective cover, provided the applied layer is of sufficient thickness. Under the mulch, the extremes of the micro-climate are eliminated, but due to the short lengths of the chopped straw, its effectiveness in extreme, difficult sites is usually limited.

Advantages
Mechanical operation and rapid work progress at low cost on even slopes at accessible sites.

Disadvantages
Only feasible at easily accessible sites, very limited scope of application on slopes with a gradient steeper than 1:1, and under difficult site conditions. Limited reach. Because of the high cost of the machinery, its use on areas of less than one hectare is, as a rule, uneconomical.

Areas of use
Shallow slopes of great length.

Mulch seeding using full-length straw (the Schiechteln method)

Materials
10–50 g of seed per sq m, depending upon site and purpose; 300–700 g per sq m of long-stem straw or hay, or structurally similar organic or artificial fibres; 40–60 g per sq m of inorganic fertiliser or 100–150 g of organic manure; 0.25 l per sq m of bitumen emulsion; 0.25 l per sq m of water; sundry technical and biological preparations, depending upon site conditions.

Implementation
As a first stage, the straw is spread on the slope, resulting in an even layer of mulch. Depending upon the site, the straw is pre-treated employing different methods. In the second stage, the treated seeds, mixed according to site requirements and objective, are evenly spread, at the same time applying the fertiliser or manure. If necessary, soil improvers, soil stabilising agents or growth promoting substances are added. During the third stage, the mulch is fixed on the soil surface to prevent it shifting. This is achieved by spraying the mulch with a specially prepared, plant compatible, diluted but stable, bitumen emulsion. On sites where the use of bitumen is prohibited (catchments for town water supply, etc.), or where the dark colour of the emulsion is not wanted, different bonding agents may be used. All work phases may be carried out by the use of machinery or by hand labour. Daily work output varies between 3000 and 15 000 sq m per application, depending upon site conditions. If the application of the binding emulsion is insufficient to secure the mulch in place, protruding stakes or nails may be hammered into the soil prior to spreading. For this purpose 350 mm lengths of reinforcement steel are best, aiming at a density of one peg per sq m. If necessary, the mulch layer may be further secured by wires, securely anchored to the stakes.

Timing
During the growing period.

Effectiveness
The nature of the long straw mulch must be such that, depending upon local conditions of temperature, precipitation and daylight, light penetration is maintained. Under the straw mulch, a sufficiently large volume of air is available to enable a climatic buffer zone to be created; this air warms up very quickly without reaching a level which is injurious to plant life. Dehydration is slowed down, and radiation losses during the night will cause condensation. The stimulating effect this has on the germinating plants is quickly seen. Close observation of control areas has clearly shown that plant growth is repeatedly checked during the course of the average day; this does not happen when the Schiechteln method is employed. As a consequence, plant mass production is appreciably increased per unit time, resulting, under favourable conditions, in one month's growth exceeding that achieved by unmulched areas over an entire growing season. The dark colour of the bituminous binder is of advantage, as it causes rapid warming which leads to accelerated germination. This effect is highly beneficial, particularly in

mountainous terrain and/or when planting takes place towards the end of the growing season. At different sites, this may be a disadvantage and light coloured emulsions are to be preferred. As already mentioned, the mulch not only influences the micro-climatic conditions, but also protects against mechanical forces such as rockfalls, hail, raindrop impact, wind, etc.

Advantages
Simple, quick-acting method at reasonable cost. No special access to the site is required. With the exception of the bitumen application, all work can be done by hand. There are, however, many types of machinery available to do the job. The method is specially suited for sites in difficult terrain and under adverse soil conditions and the final result is of a high standard as all critical work is done by hand. The most labour intensive part of the operation – the spreading of the mulch – can be done by unskilled labour.

Costs
These are dependent upon the site conditions and size of area to be planted, general access, soil conditions and the type of cover envisaged, (cost of materials). On average, costs amount to 0.06–0.2 work hours per sq m.

3.1.3 Direct seeding of shrubs and trees

This method has two important areas of application:

- at difficult sites, where normal planting of trees and shrubs is precluded due to stony, rocky or precipitous terrain;
- as a supplementary measure on difficult ground where, after the completion of all works, patches of poor or zero cover are evident. Suitable seed varieties are listed in Table 2.2.

Materials
Seeds of broadleaved and coniferous trees and of shrubs. The seeding rate depends upon the size of the individual seed and the germination percentage. Small seeds may be mixed with sand, sawdust or similar material to avoid waste. The quality of seed is controlled by the relevant regulations in force and laid down by the relevant authorities. The origin of the seed is just as important as the origin of planting material. Prior to sowing on raw or mineral subsoil, the seed should be treated with the appropriate root fungus inoculants.

Implementation
Several methods, based on forestry practice, can be considered for use:

- *Broadcast:* the spreading of seed by hand or by machinery onto a prepared or natural seed bed. This method is particularly adapted for sowing small or light seed. In principle, the same methods are employed as those used for the seeding of grassland, namely hydroseeding (see Section 3.1.2.3). The seed mix (grasses, herbs, woody plants), together with the usual additives such as fertiliser and binding agents, etc., is emulsified in water and sprayed onto the soil surface in one operation.
- *Hole planting:* holes of approximately 100 mm depth and width are dug with a hoe or similar implement. One to five large seeds, as for example acorns, are placed in the hole, or a little amount of smaller seed is dribbled into the planting station. The seed is covered with 10–20 mm of fine soil.
- *Spot seeding:* seed is scattered on locally prepared small patches varying in size from 20–40 sq mm to 2–4 sq m. The existing vegetation is chopped down, the soil surface loosened, and the seed applied.
- *Drilled seed:* the seed is placed in prepared shallow furrows, either by hand or with the use of a seed drill, and lightly covered with earth.

Timing
Seeding is best carried out in spring or autumn. In spring after snow melt, the soil should be at field capacity for quick germination. Seeds sown in autumn, which resembles the natural course of events, lie dormant during the winter months. The small amount of artificially sown seed, as compared with the vast quantities available annually from the natural vegetation, needs special protection against wild animals. Several products for this purpose are commercially on offer. In the past, snow seeding, i.e. spreading the seeds of larch, birch and alder onto the snow, was a common practice.

Effectiveness
Under cover of recently established vegetation, seeds of trees and shrubs can develop under conditions resembling those of natural succession. The diversity of woody plants is increased and, with it, the diversity of future plant associations.

Advantages
Simple and low cost method. Root development will adapt to site conditions. No interference with the soil and therefore no danger of erosion.

Effective selection due to the high number of individual seeds and varied species composition as compared with the planting of individual saplings, etc.

Disadvantages
Procurement of guaranteed site-adapted seed is difficult.

Areas of use
No other planting system can be substituted for the seeding of woody plants on rocky and precipitous slopes. It is economical, and achieves good results with regard to plant selection, diversity and stabilisation by virtue of strong root penetration.

3.1.4 Erosion control nets (Plate 12)

Each of the above seeding methods may be reinforced by the additional placing of nets made from jute, coir matting, synthetic fibre or wire. Such nets fix and anchor the mulch, achieving better protection of the ground surface. To justify the considerably higher cost of additional netting, adequate reasons for its use must be established, as for example the danger of serious erosion due to extreme site conditions. Netting is justified on slopes consisting of highly erodible material (non-cohesive sands) and steep slopes exposed to wind.

3.1.5 Seed mats (Fig. 3.2)

Materials
Finished mats of variable construction based on organic biodegradeable and durable synthetic fibres; they consist usually of two different fibre layers, separated by reinforcement.

Implementation
Seed mats should be placed on surfaces which are moist and have a fine tilth. Soil surface gravel or rubble must be covered with a layer of fine material. There is no need to roughen the soil surface. After placing the mats, they must be rolled or pressed down to establish close contact with the ground surface. To prevent the mats from shifting, steel pegs or stakes are driven approximately 300 mm into the ground, or their edges are bent over and buried to a similar depth.

Ground Bioengineering Techniques 61

Fig. 3.2 Placing of seed mats. (Courtesy: Aquasol, Vienna).

Timing
During the growing period.

Effectiveness
Long-lasting and good cover provided as the fibres take a long time to break down. Limited protection against mechanical forces and often low water storage capacity.

Advantages
Mats last a long time and resist erosion as long as there is no undermining by water flowing under them.

Disadvantages
Suitable only for even ground with a fine tilth. At the present time, there are only standard seed mixes available. Special mixes must be ordered well in advance of use.

Costs
Depend upon make of the mat and type of slope to be protected. Usually higher than normal grass seeding.

Areas of use
Predominantly shallow water ways where only temporary flooding occurs.

3.1.6 Precast concrete cellular blocks (Fig. 3.3 and Plate 13)

Fig. 3.3 The selection of a concrete block with larger apertures would facilitate better grass establishment.

Materials
Cellular concrete blocks, not reinforced, are available in a variety of types and under a range of proprietary names; anchors or reinforcement steel pegs, one per sq m; good topsoil for the fill; seeds.

Implementation
Precast concrete cellular blocks are placed on the slope surface, similar to a simple grating, and fixed with iron pegs or anchors. The voids of the blocks are filled with topsoil which is seeded by any of the previously mentioned methods.

Effectiveness
Very good, long-lasting cover; depending upon the shape of the block, the grassing effect can be very variable.

Advantages
Immediate stabilising effect. Use of manufactured blocks, freely available from the trade.

Disadvantages
Only few of the presently available blocks are designed for the purpose of

slope stabilisation; after filling the blocks with soil, exposed concrete is unsightly for some time. High cost.

Areas of use
Protection of unstable slope sections, specially at the slope base which must provide a secure foundation for the blocks; protection of riverbanks.

3.1.7 Live brush mats (Fig. 3.4)

Materials
Live willow stem cuttings and branches; 20–50 stems per running m.

Implementation
The slope surface is covered with live stem cuttings and branches laid butt end downslope so as to achieve a minimum of 80% cover. The lower, thicker ends of the branches must be covered with soil to facilitate rooting. In addition, their butt ends must be fixed by stones, fascines, poles or wattle fences. If the lengths of the cuttings are shorter than the slope, several must be laid in line, with overlaps of at least 0.3 m. To ensure the proper rooting of the mat, close ground contact and a fine surface tilth are essential. Finally, a light soil cover is applied and the mats fixed to the slope with stakes and wires (or by stems pinned down transversely at regular intervals up the slope).

Timing
Only during the dormancy period.

Effectiveness
The mats provide an immediate cover and prevent erosion. Intensive stem growth and root development will follow with the onset of warm weather.

Disadvantages
Large masses of live material required. Labour-intensive construction method.

Costs
Depending upon slope characteristics and the outlay for the branches, 1–5 work hours per sq m.

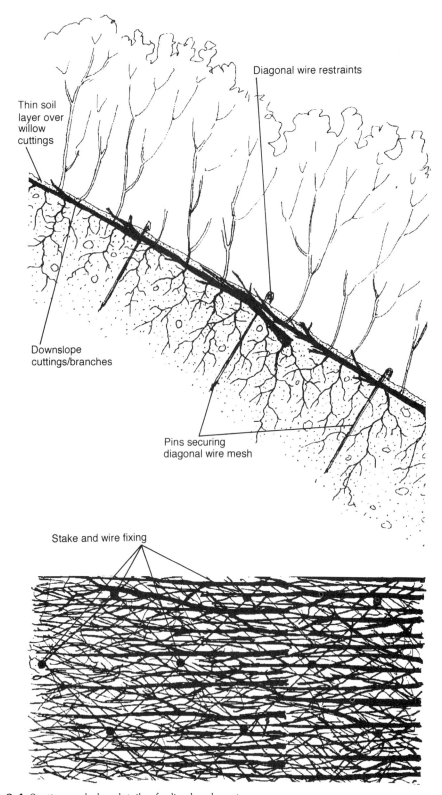

Fig. 3.4 Section and plan details of a live brush mat.

Areas of use
For the rapid protection of slopes against erosion caused by surface run-off and wind.

3.2 Ground stabilising techniques
(Figs 3.5–3.17; Plates 14–27; Table 3.4 near the end of this chapter)

Ground stabilising techniques are adopted wherever adverse subsurface mechanical forces threaten the stability of earth structure and its surroundings. The immediate effectiveness of the described measures depends on how deeply they penetrate the ground and the distance between them. The spreading root system enhances their effectiveness which increases over time and with the growth rates of individual root structures.

Ground stabilising techniques are always linear or fixed point systems that require supplementing with effective soil protection works. (For further information on the stabilisation of slopes based on vegetative methods, see Schaarschmidt and Konecny, 1971.)

3.2.1 Live cuttings (Fig. 3.5 and Plate 14)

Materials
Cuttings – i.e., unbranched and healthy one-year-old or older stems of suitable plants of a diameter of 10–50 mm and a minimum length of 400 mm (see Table 2.3).

Implementation
Planting of cuttings: a crowbar or similar metal rod of suitable diameter is used to punch narrow holes into the ground into which the cutting is placed; the soil is then firmed down around it. There is no harm in hammering the cutting a little further into fairly soft soil, provided the basal cut is on the slant; even mechanical hammers operated by compressed air may be used for the planting of substantial cuttings. A short section of pipe of a suitable diameter is welded onto the tip of the chisel to prevent the hammer from slipping off the top of the stake. Only about one quarter of the cutting is to protrude above the soil surface, to prevent it from drying. The cuttings should be placed at random at the most suitable places, never in rows. If possible, at least two cuttings should be planted per sq m, preferably five in more exposed locations.

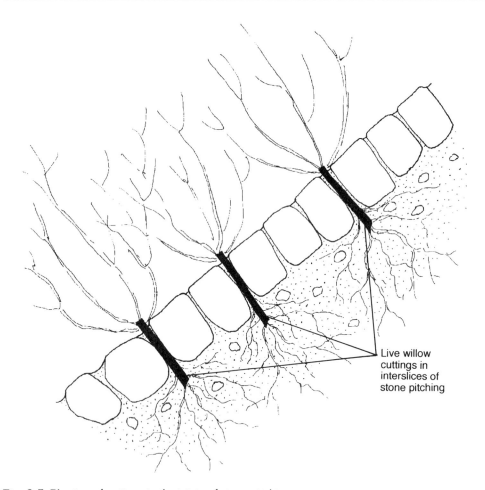

Fig. 3.5 Planting of cuttings in the joints of stone pitching.

Planting into joints

Cuttings are driven into the gaps of dry stone walls to a depth which embeds them firmly into the ground behind the wall. The insertion of the cutting is again facilitated by the use of a pointed metal bar. After the cuttings have been placed, the hole should be filled with dry sand, or preferably with soil, which does not have to be topsoil. The number of cuttings planted per sq m depends upon the size of the stones or blocks used (the smaller the stones the greater the number of cuttings). In a dry habitat, cuttings placed in stone pitching grow better than those planted into the open ground. An average failure rate of 30–50% can be expected.

Timing
Only during dormancy.

Effectiveness
Stabilising effects occur only after root development; at the same time, a certain drainage effect is achieved by the water requirement of the plants. The developing root system binds the facing tightly into the ground, so that for this system, smaller stones may be used. The beneficial effect such joint plantings have on local site conditions is comparable with that expected from the establishment of bush by other planting methods using branches capable of vegetative propagation.

Advantages
Low-cost ground stabilising technique and rapid work progress. Can be used on existing walls. Smaller stones can be used for constructing the facing.

Costs
Very low, 0.05–0.1 work hours per cutting. Planting of cuttings on slopes: 5–7 sq m per hour; planting into joints: 2–4 sq m per hour, including collection of cuttings.

Areas of use
Large scope in all aspects of vegetative protection measures, specially for the quick and low-cost planting of wet slope sections. Planting cuttings in the joints of dry stone walls for reinforcement.

3.2.2 Wattle fences (Figs 3.6, 3.7 and Plate 15)

Materials
Long, pliable willow whips or stems, capable of throwing shoots.

Construction
Wooden stakes, 30–100 mm in diameter, or steel pegs 1 m long, are hammered into the soil at 1 m intervals. Another shorter stake is driven into the soil halfway between two pegs. Strong, flexible live stems are then plaited or woven between the stakes, each pair of plaits being pressed firmly in turn into the ground. Three to seven pairs of cuttings are plaited over each other in this way. Alternatively, instead of individual cuttings, prepared wattle panels may be fixed to the stakes. The stakes should not protrude more than 50 mm above the wattle with not less than two-thirds of their length being firmly embedded in the soil. At least the lowest stem cutting and the butt ends of all others must be

68 Ground Bioengineering Techniques

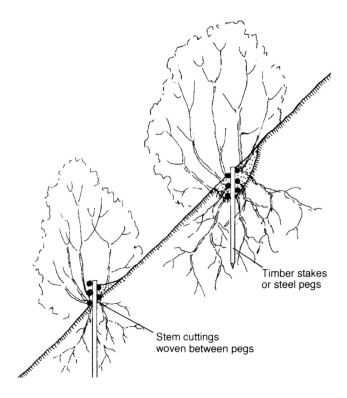

Fig. 3.6 Buried (*top*) and exposed (*bottom*) wattle fences.

buried in the soil to enable them to take root. Completely soil-covered wattle fences have a better chance of taking than those being placed above the surface. Exposed cuttings may dry out and die. Wattle fences may be aligned in straight lines or in a diagonal pattern to form a rhombus or diamond shape.

Timing
Only fences constructed during the dormancy period have a chance of survival.

Advantages
Permits immediate ground stabilisation, and enables stable terraces to be formed.

Disadvantages
- Very high demand for live plant material and relatively low rooting percentage, as many stems are above ground level. Only long and

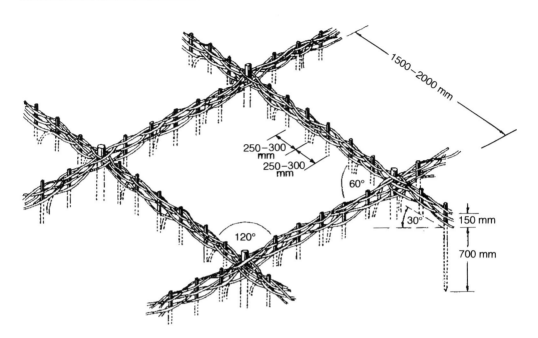

Fig. 3.7 Isometric plan view of a diamond-shape wattle fence.

pliable cuttings are suitable, which precludes, particularly in the Alps, many valuable and well-adapted willows.
- Horizontally arranged wattle fences have all the cuttings at right-angles to forces exerted by rockfalls and slides. A single stake must therefore absorb the total force exerted by one fence section: if it fails, the whole fence is in danger of collapsing. Individual stakes are easily broken by rockfalls, or levered out of the soil by snow pressure.
- High cost and work-intensive construction method.

Costs
0.8–1.5 work hours per linear m.

Areas of use
Wattle fences should always be fully sunken into the ground. There will always be a place for them, e.g. to stabilise small slides or to retain re-applied topsoil, and in combination with other construction methods.

3.2.3 Fascines (Fig. 3.8 and Plate 16)

Materials
Preferably straight and long branches of shrubs and trees capable of vegetative propagation.

Construction
Fascines of live branches are placed into ditches 0.3 m wide and up to 0.5 m deep. Satisfactory results will be obtained if at any given section of the fascine, five stems of a minimum diameter of 50 mm are incorporated. These relatively thin fascines, as compared with those used in hydraulic engineering works, save construction material; they are buried at a shallow depth and have the added advantage that most, if not all, of

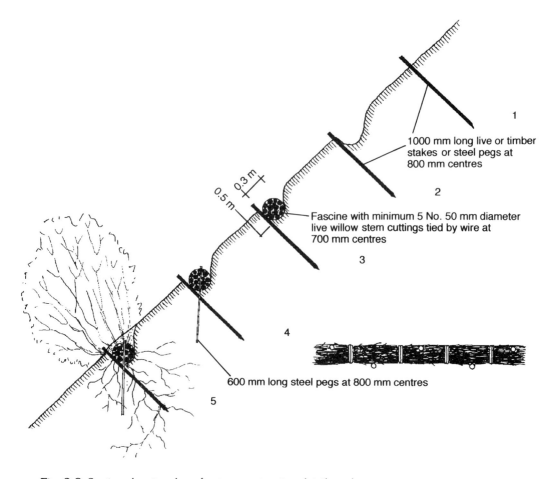

Fig. 3.8 Section showing slope fascine construction details and sequence.

the embedded stems are in contact with the soil and therefore have a chance to take root. There is no need to tie fascines as tightly as those used in hydro engineering (or riverworks). It is sufficient when the ties are 700 mm apart. The fascines are fixed in the trenches by live or dead wooden stakes or steel pegs which are some 600 mm long, at 800 mm intervals. The stakes must be driven vertically into the soil, their upper ends being level with the top of the fascine. Formerly, the stakes or pegs used to secure the fascine were placed just downslope of the fascine. In more recent times, the pegs are driven through its centre, thus saving binding wire. Owing to their small cross-section, steel pegs are better suited for this purpose than wooden stakes. Immediately after placing the fascines, the trenches are backfilled, leaving only short sections of the cuttings to protrude above the soil.

Timing
Only during the dormancy period.

Effectiveness
Retains water if aligned horizontally and assists drainage if on a gradient (fascine drains). Soil stabilisation effect only after rooting.

Advantages
Simple and effective stabilisation structure, short construction time, and little earth movement involved.

Disadvantages
Only slim, relatively unbranched stem cuttings can be used. Stabilising effect confined to shallow depth. Sensitive to rockfalls.

Costs
0.5–1 work hours per linear m.

Areas of use
Very effective on cut slopes, in deep and soft soil in climatically favourable areas.

3.2.4 Fascine drains (Fig. 3.9 and Plate 17)

Materials
Long and straight live stems or branches combined with dead branches – recently cut so as not to be brittle.

72 Ground Bioengineering Techniques

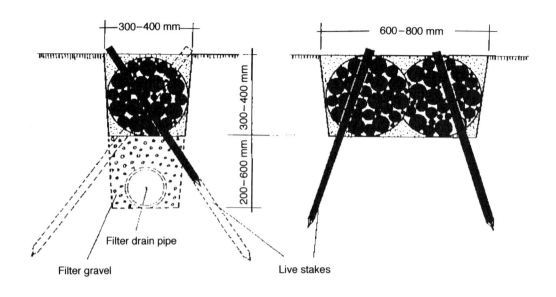

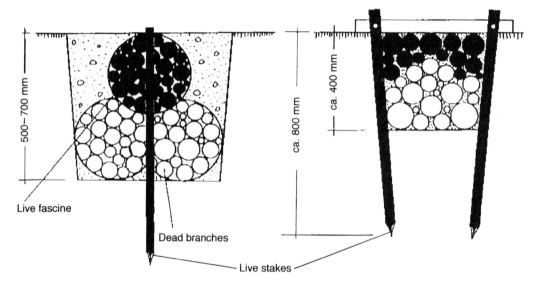

Fig. 3.9 Various types of fascine drains.

Implementation

Fascines of the required length and 200–400 mm diameter are constructed, using live and/or dead branches, the cut ends always facing the same way; 1.5–3 mm thick wire is used for the ties, which are 500 mm apart. The fascine is placed into the ditch, filling it completely. If required, several fascines may be placed on top of each other. The lower fascines are made of dead branches which must be green and fresh. If drainage to a greater depth than 300–400 mm is required, deeper ditches are excavated and partly backfilled with filter gravel, leaving enough room for the fascines to lie flush with the soil surface. The drain is then backfilled with earth, covering all branches completely to facilitate rooting. The fascines are firmly anchored by driving live stakes of approximately 50 mm diameter and 600 mm length every 800 mm at an angle through them. The fascine drain discharges into the main drainage system. The alignment of such drains is usually straight downslope. To prevent the fascine from being torn apart on very steep slopes, strong wire or wire rope may be incorporated to resist the force of gravity. The wires or ropes are then anchored firmly at the top of the slope. For the effective drainage of long slopes, several drains running parallel to each other at a distance of 1.5–3 m may be installed, reinforced as necessary. Stärk (1963) has developed 'reinforced cabled willows' which were successfully used in particularly difficult terrain during the construction of the West Autobahn near Vienna.

Timing
During the dormancy period.

Effectiveness
Fascine drains are immediately effective, ensuring unimpeded flow along the parallel placed branches. After the branches have rooted and the shrubs grown up, their drainage function is increased by the water use of the vegetative cover.

Advantages
Simple and fast construction, still effective after the original ditch is silted up. Cheaper and more attractive than conventional methods on slopes that need drainage only to a depth of 400 mm. Suitable for area drainage.

Disadvantages
Only long stems and branches can be used. Large quantities of plant material necessary. Labour-intensive.

Costs
1–3 work hours per linear m.

Areas of use
Drainage of natural slopes and embankments.

3.2.5 Furrow planting (Fig. 3.10 and Plate 18)

Materials

Live fascines consisting of three to ten branches of live shrub cuttings or young rooted shrubs, stakes, topsoil or compost at a rate of 0.05 cu m per linear m.

Construction

Furrows 300–600 mm wide and approximately 250 mm deep are dug across the slope. Thin and live fascines are placed at the downward side of the furrow and fixed by driving stakes through them at 1 m intervals. In the upslope side of the furrow, young, rooted plants of suitable shrubs or trees are planted some 500–1000 mm apart. The furrow is backfilled with topsoil, or a topsoil–manure mix. Such furrows may impede the free

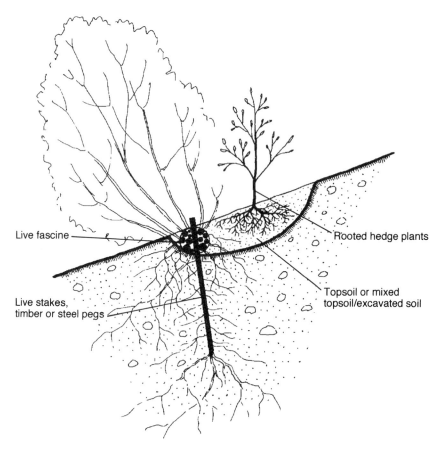

Fig. 3.10 Furrow or groove slope drain.

flow of surface and/or subsurface water which may cause slides. It is therefore advisable to angle the furrows at 10–30° from the horizontal, or arrange them in a herringbone fashion. The actual angle depends on the nature of the subsoil and on rainfall intensity. On permeable soils, a flat angle is preferred; on very heavy and rather impermeable soils, the construction of furrows may be inappropriate, as retained water may cause damage.

Timing
Only during the dormancy period.

Effectiveness
If aligned at an angle, the furrow acts as a drain. The rooting fascine will stabilise the soil. Due to topsoil and manure backfill, trees and bushes planted in the furrow will grow faster and better than those on the untreated slope.

Advantages
Combined effect of drainage and water conservation. Correct plant selection allows for the establishment of pioneer and climax vegetation.

Disadvantages
Labour-intensive. Only on gentle slopes up to a maximum of 30°. Not suitable for rocky conditions. Topsoil required.

Costs
1–3 work hours per linear m.

Areas of use
On flat, marshy slopes in climatically favourable areas.

3.2.6 Cordon construction (Praxl, 1961) (Fig. 3.11 and Plate 19)

Materials
Live cuttings; poles; conifer branches.

Construction
Horizontal terraces, 0.5–1.5 m wide and 1–3 m apart are cut into the slope, angling the surface backwards at a 10° inclination into the slope. Long round poles are placed onto the terrace which are covered with conifer branches. These are in turn covered to a depth of 100 mm with

76 Ground Bioengineering Techniques

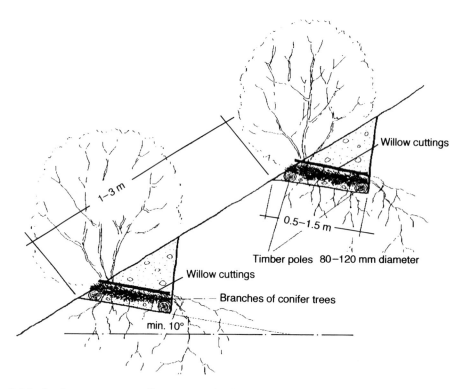

Fig. 3.11 Cordon construction. (Source: Praxl, 1961)

soil; live willow cuttings, 20–30 mm apart, are then placed on the soil and covered with backfill.

Timing
Only during dormancy.

Effectiveness
Effective stabilisation of slope sections prone to slides; quick rooting of cuttings due to loosening of the soil during construction.

Advantages
Effective ground stabilisation enhanced by the flexible but firm layer of conifer branches. Easy rooting in the loose soil of the completed cordon.

Disadvantages
Labour-intensive construction. Large quantities of conifer branches required.

Costs
Most expensive of the stabilisation methods; 3–4 work hours per running m.

Areas of use
On steep slopes with heavy-textured, easily saturated soil.

3.2.7 Layering (Figs 3.12–3.15; Plates 20–25)

Layer construction methods are subdivided into three variants which have developed over the years:

- hedge layers using rooted plants;
- brush layers using sprouting branches;
- hedge–brush layers using rooted plants and live branches/brushwood.

The criteria for the preparatory work required for the above-mentioned three types of construction methods employed in the stabilisation and protection of natural slopes in general, and for cuttings and embankments in particular, are the same. Terraces or berms of between 0.5 and 2 m width and approximately equal depth are cut into the slope; on very steep sites, the berms are replaced by 0.5 m deep trenches. Construction starts at the toe of the slope and, as the work progresses upslope, soil excavated from the adjacent higher terrace is cast downslope to cover the brush layer on the one below it. The base of the berm should slope at an angle of 5–10° backwards to provide a secure bed for the brush layer. The terraces are either built on the contour, or, if surface or subsurface water is to be disposed of, angled; the deviation from the horizontal should not exceed 60°. Steeper angles will complicate construction. The distance between two berms depends upon slope inclination and soil conditions, but varies on average between 1 and 3. Closer spacing than 1 m is not practical because of the danger of large soil sections breaking away and collapsing onto the lower terrace or trench. In small areas, layer berms or trenches could be constructed by hand; otherwise small crawler equipment, hydraulic excavators or cable-drawn reversible ploughs are used.

3.2.7.1 Hedge layers (Fig. 3.12)

Materials
Rooted plants of deciduous shrubs or trees which tolerate complete soil cover and produce vigorous adventitious roots (see Table 2.3). By

78 Ground Bioengineering Techniques

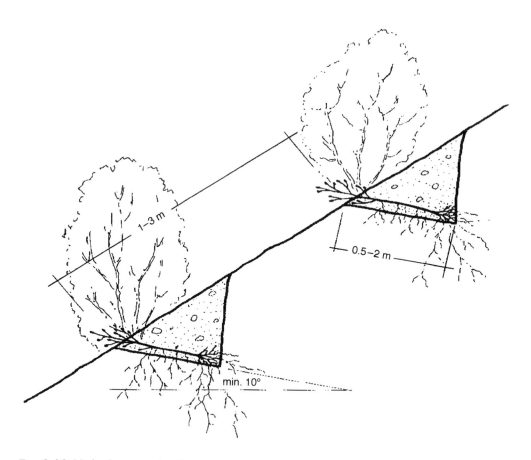

Fig. 3.12 Hedge layer construction.

preference, two- to four-year-old strong saplings should be used; fast growing species, for example alder, are best planted as two-year-old stock. The ratio of stem and branch wood to root is of importance. Plants with a well-developed, massive root system establish themselves more easily and grow more vigorously. Depending upon the species, 5–20 plants per linear m are required.

Construction
Preparation of the berms of 0.5–0.75 m width and depth. To improve the soil, a thin layer of topsoil or organic matter such as manure or straw may be spread on the berm. The rooted plants are laid flat on the berm base, close to each other with one-third of their length protruding beyond the surface of the slope.

Timing
Spring or autumn.

Effectiveness
It is possible to establish a deciduous woodland association without first preparing the ground by other cultural measures. Initial plant selection should include species intended for the appropriate climate type.

Advantages
Telescoping and shortening the process of plant succession, by commencing at mid-succession.

Disadvantages
High demand for plant material. Only suitable at favourable site conditions.

Costs
1–3 work hours per running m.

Areas of use
At favourable sites with fertile, good soil, where willows are rare or do not occur at all, or where willows are unobtainable.

3.2.7.2 Brush layers (Figs. 3.13 and 3.14; Plates 20 and 21)

Materials
Exclusively branches and parts of branches of deciduous woody plants capable of throwing shoots, predominantly willows (see Table 2.3). Twenty branches per linear m.

Construction in cuttings and non-compacted fill
The branches are placed very closely in a criss-cross manner on the berm which should be between 0.5 and 2 m wide and of similar depth. Care should be taken to place the branches at random with regard to size and age and species composition. After replacing the soil, the branches should, irrespective of length, protrude 0.25 m beyond the surface of the slope.

Construction in compacted fill
It is possible to establish the brush layer during the construction of embankments and dam walls. Large branches of 2–5 m length are placed

80 Ground Bioengineering Techniques

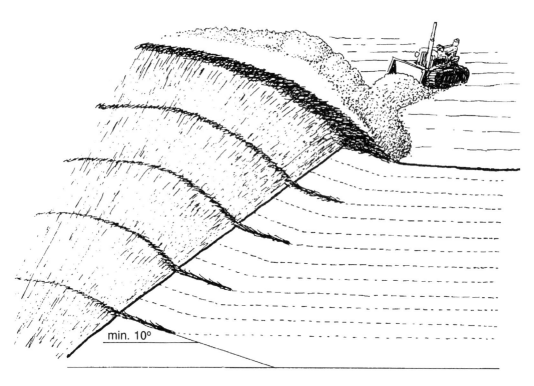

Fig. 3.13 Synchronous brush layer construction in embankments.

in crossed pairs on top of the soil at the chosen spacing and covered with fill material which is compacted to specification. This does not damage the brush layer which should be tilted between 5 and 10° into the wall in the usual way.

Timing
Only in the dormancy period.

Effectiveness
Among the ground stabilising techniques, the brush layer has an immediate impact, its protective and stabilising effect extending into lower soil horizons. At extreme sites where erosion, deposition and rockfall are particular hazards, brush layers and the pioneer vegetation that develops with them are gradually eliminating these problems.

Advantages
Relatively simple construction. Fast establishment of a stable soil-root complex. Relatively short and spreading branches of the scrub willows

Ground Bioengineering Techniques 81

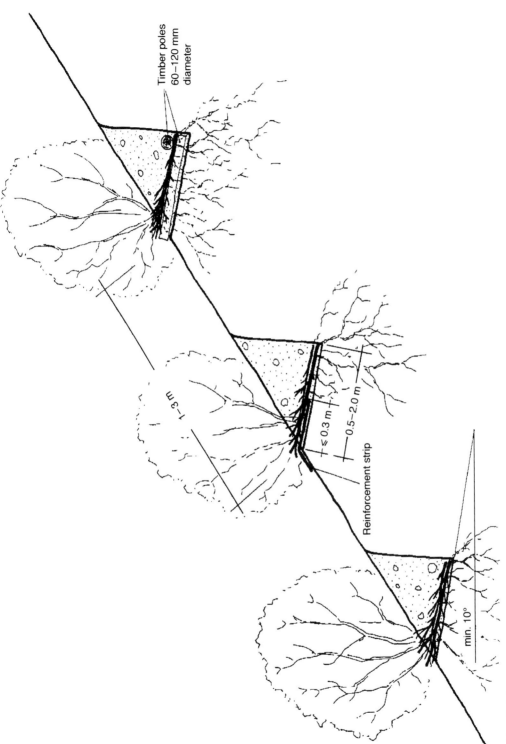

Fig. 3.14 Brush layer construction in cuttings and on slopes prone to shallow slips. Standard version (*bottom*); brush layer with geotextile, geogrid, felt or d.p.c. layer reinforcement strip (*centre*); massive brush layer, reinforced with poles (*top*).

growing in the subalpine region can be used. Simultaneous construction during fill operations possible.

Disadvantages
Not suitable for the stabilisation of deep, organic topsoil layers.

Costs
0.7–2 work hours per linear m.

Areas of use
Stabilisation of land slides and hill creep, torrent control and avalanche protection.

3.2.7.3 Hedge–brush layers (Fig. 3.15; Plates 22–25)

Materials
Live branches (see 'Brush layers') and bare rooted saplings are planted together: 10 branches and 1–5 saplings per linear m.

Construction
Branches and saplings are placed alternately onto the berm, with plants and branches protruding 200–300 mm from the slope surface.

Timing
Only during dormancy.

Effectiveness
The use of rooted plants together with live branches shortens the natural succession. Accelerated improvement in site conditions as compared with brush layers.

Advantages
Establishment of several age groups in one operation.

Disadvantages
Limited suitability for the stabilisation and conservation of organic topsoil.

Costs
0.8–2.5 work hours per linear m.

Areas of use
Compared with simple brush layers, hedge–brush layers offer greater

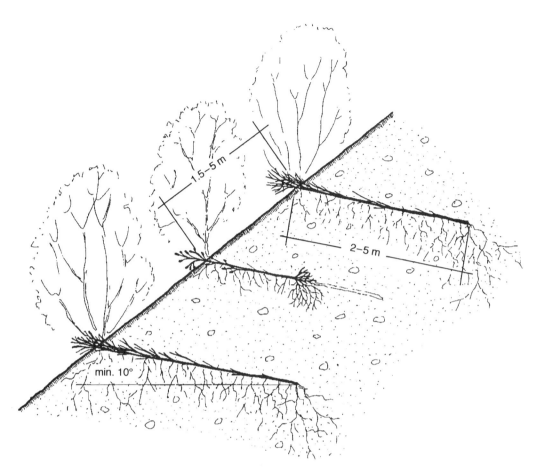

Fig. 3.15 Hedge–brush layer.

versatility for the protection and stabilisation of natural slopes, cuttings and embankments below the timber line, permitting the combination of shrub and tree plantings. All layer construction methods are sub-divided into two versions:

❑ On dry slopes and fine grained material, the first method uses so-called reinforcement strips made of planks, sheet metal, bituminated roofing paper, plastic foil, geotextiles, geogrids, etc. These are 100–300 mm wide strips which are placed longitudinally at the edge of the berm below the brush layer. Their principal function is moisture conservation, prevention of erosion at the edge of the terraces and reinforcement of the backfill.

❑ On extremely steep slopes prone to slides, the second version uses

84   Ground Bioengineering Techniques

wooden poles up to 150 mm in diameter, placed across or along each berm/terrace either above or below the brush layer. This prevents dislocation of the layers due to slump or creep.

3.2.8 Gully control (Fig. 3.16; Plates 26 and 27)

Materials
Live branches capable of vegetative propagation.

Construction
To achieve a dense cover and intensive rooting, the branches are placed in a herringbone arrangement in the gully, with the cut ends at the bottom of the gully. Each branch layer is covered with soil, leaving the tips free, taking care that total soil cover at the base does not exceed 0.5 m. To keep the branches firmly in place, cross beams or poles are securely embedded every 2 m in the sides of the gully. These structures are fairly resistant to periodic cycles of erosion and deposition.

Timing
During the dormancy period.

Effectiveness
The intensive root development protects the sides and the bottom of the gully.

Advantages
Permanent protection due to the use of live material.

Disadvantages
Large quantities of live branches required.

Costs
Relatively low, provided live branches are available in close proximity to the construction site.

Areas of use
Rehabilitation of up to 3 m deep erosion channels and gullies subject to periodic flooding. Efficient control of slow but steady erosion.

Ground Bioengineering Techniques 85

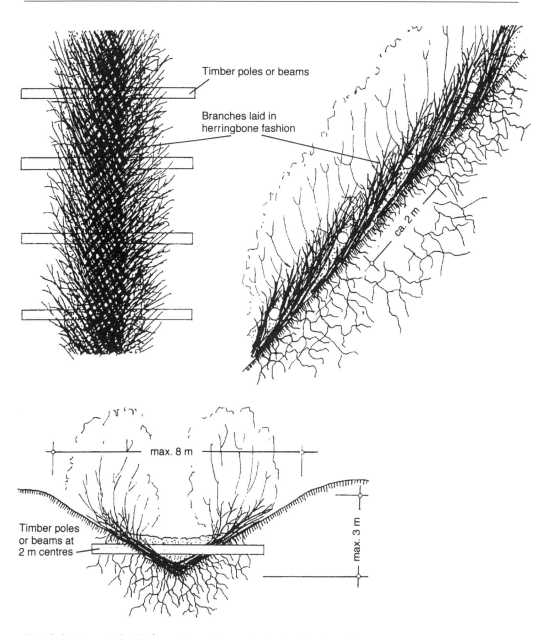

Fig. 3.16 Layout details for gully erosion control using live branches.

86   Ground Bioengineering Techniques

3.2.9 Stake fences (Fig. 3.17)

Construction
Live, relatively uniform stakes, cut obliquely square at upper ends, are hammered for a third of their length in close proximity to each other into the ground, tied securely with steel wire or willow stems to cross beams which are keyed deeply into the sides of the gully.

Timing
Only during dormancy period.

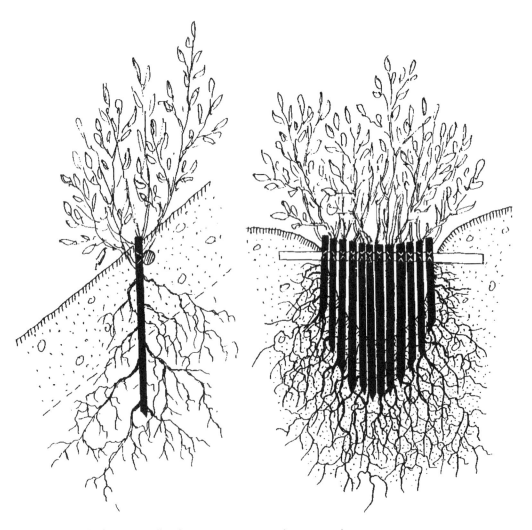

Fig. 3.17 Stake fence or pallisade construction using live materials.

Effectiveness
The stake fence becomes immediately effective, even before rooting has taken place, as it provides a barrier that protects the bed and causes siltation; this effect is increased after rooting and sprouting of the stakes has taken place.

Advantages
Fast construction, immediately effective and forms dense vegetative cover. Simple and effective method to establish checks in steep and deep gullies using live material.

Disadvantages
Limited to gullies approximately 5 m wide and 2–4 m deep. The stakes should be straight and the stout poles up to several metres in length: this confines the stake fence technique to areas where growing conditions are above average.

Costs
Relatively cheap if stakes are available in close proximity to the site.

Areas of use
For construction of check dams in deep and steep gullies in climatically favourable areas, with good fine-grained soil (clay, loam, loess or sand).

3.3 Combined construction techniques
(Figs 3.18–3.25; Plates 28–45; Table 3.5 near the end of this chapter)

Structures for the control of erosion, the protection of slopes and the stabilisation of gullies are not necessarily always made of inert building materials only, but they may, to advantage, be combined with live plants or parts of plants. Such structures become immediately effective upon completion. As time passes, increasing root mass and vegetative cover enhance the effectiveness of the measures adopted. As a rule, combined construction techniques can be programmed to provide rapid stabilisation ahead of other measures which use live materials only.

3.3.1 Vegetated dry stone block walls, stone pitching and revetments (Plates 28–30)

Materials and timing
Live branches and bare-rooted plants may be only employed during the dormancy period. Sods and turf sections may, with the exception of frost

periods, be planted throughout the year; mulch seeding can be done only during the growing period. The combination of several plant types and methods may, depending upon site conditions, be of advantage.

Construction
During the construction of dry stone block walls, stone pitching, rock revetments, etc. live branches or bare-rooted shrubs are placed in the joints and firmly embedded into the earth behind the stones, penetrating any filter gravel. Live branches should not protrude for more than 0.3 m into the air, to prevent desiccation. For best results, after the completion of the wall, the branches are cut flush with its surface. If for ecological reasons shrubs and trees are out of the question, turf may be packed into the joints. The use of sods in conjunction with low dry stone walls goes back for many years; it is, however, important that the face of the wall is well-battered. It is possible to fill the joints of such moderately inclined stone walls or stone pitching with earth after completion in preparation for mulch seeding. For this, topsoil is not necessary, but there must be sufficient fine material in the joints to permit rooting.

Advantages
Use of stones of all sizes, flexibility, permeability, low cost and long life. A further advantage compared with concrete walls or cement grouted stone walls is that in the event of the wall being demolished, the stones can be reused. Grassed or planted walls are usually more attractive than bare walls.

Disadvantages
Live branches and rooted shrubs and trees can only be planted during the dormancy period. Total height is limited.

3.3.2 Filter wedge (Fig. 3.18; Plate 31)

Construction
Permeable material such as gravel is placed in layers at the base of the slope. During construction, live branches are placed singly or in layers into the filter, so that the cut ends reach the soil behind. Planting after the completion of the wedge is only possible if the type of material used and the thickness of the filter permit penetration; grassing can take place at the same time. Filter wedges are effective for the stabilisation of large slides, by placing a layer of very coarse gravel, 0.6–3 m thick, between the

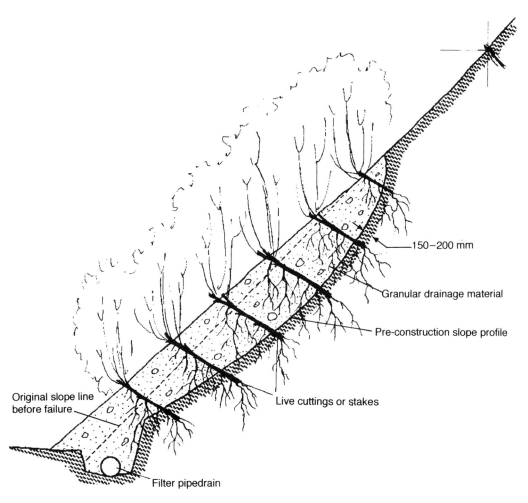

Fig. 3.18 Typical section showing filter wedge construction.

outer filter and the saturated slope. The supporting gravel layer may be reinforced by a series of very strong live stakes and/or brush layers, which must reach the soil behind the filter.

Timing
During the dormancy period.

Effectiveness
Ideal combination of passive and active drainage caused by the consumptive use of water by the shrubs and trees.

Advantages
Economical and natural construction method.

3.3.3 Vegetated gabions (Figs 3.19 and 3.20; Plates 32 and 33)

Construction
Relatively fine wire mesh is placed on level ground at the site and layers of coarse gravel and stone are spread onto it, into which live branches or bare-rooted plants are inserted. To provide a proper bed for the plants, the wire netting is periodically raised and the plants inserted through the mesh. Finally, the net is drawn together and wired closed, resulting in a shape that fits exactly into the terrain. If there is any risk of movement, the gabion is fixed by means of long steel pegs which are hammered into the ground. Gabions, heavy gauge pre-fabricated wire panels forming boxes connected together and filled with large stones, are not easily planted with any kind of vegetation in the normal manner. Instead, live branches, cuttings or stakes are placed in the horizontal joints between courses of gabions.

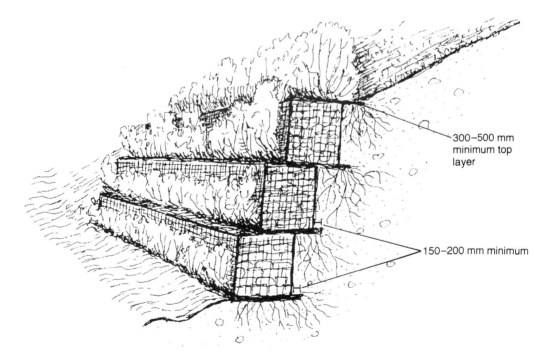

Fig. 3.19 Slope stabilisation using vegetated gabions.

Ground Bioengineering Techniques 91

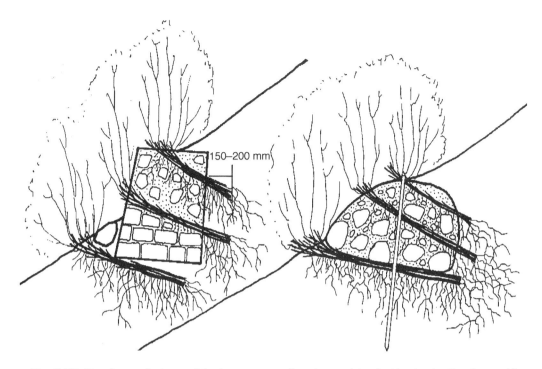

Fig. 3.20 Two forms of wire mesh basket structures (free-form gabions) with mixed soil and stone fills.

Timing
Only during dormancy period, because planting after completion is difficult, if not impossible.

Effectiveness
Provides a flexible base and support.

Advantages
Fast and simple construction. Remains flexible. Can be used as a linear retaining wall or debris barrier.

Disadvantages
Can be used only where coarse gravel or stones are available on site. Construction only during the dormancy period. Subsequent planting virtually impossible.

Costs
Economical.

3.3.4 Vegetated geotextile earth structures (Figs 3.21–3.23; Plates 34 and 35)

Materials and timing

The most effective combination with live plants is achieved if live branches or rooted plants are incorporated during construction; this limits construction to the dormancy period, but subsequent planting or insertion of cuttings is possible. Grassing and sodding, depending upon local climatic conditions, may be carried out throughout summer.

Construction

The containers, bags or sheets are made of geotextiles, a strong fabric or mesh consisting of weather-resistant synthetic fibres, termed geogrids, with the mesh aperture not exceeding 5 mm. This permits the use of fine-grained material for the fill. Three variants of this type of construction are:

- *Layer method*: The geogrid sheets are spread onto the ground from the back. Half their widths are covered with a layer of the relevant fill material to a depth of 300–500 mm which is then compacted, and the rest of the sheet is folded up and over the front of the fill and pinned in place. In this manner, any number of layers may be constructed one on top of the other to any length. For best results, brush layers or hedge–brush layers are placed during construction between the geogrid sheets.
- *Sandbags:* These are filled with gravelly soil, sand or similar material and stacked to a limited height. Live branches, rooted plants or cuttings are placed between the bags, making sure that the cut ends or

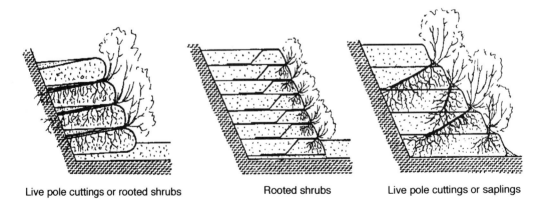

Live pole cuttings or rooted shrubs   Rooted shrubs   Live pole cuttings or saplings

Fig. 3.21 Various combinations of geotextile/geogrid earth structures with live vegetation.

roots of the plants reach the natural ground behind the bags. If the bag wall is built to any height, it is essential that long steel rods or geogrid sheets are used to stabilise and anchor the structure.

❑ *Pockets:* For localised stabilisation on steep slopes in combination with live shrubs and trees: root penetration anchors the pockets (Fig. 3.22).

All three variants may be grassed and planted with shrubs either during construction or after completion, by means of hydroseeding and/ or cuttings, container plants, stakes, etc., which are inserted through slits cut into the exposed faces of geogrids.

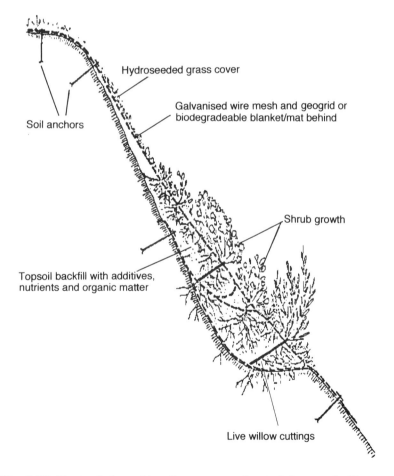

Fig. 3.22 Geotextile/geogrid earth structure under steel slope mesh. (Source: Sauli, 1986)

Effectiveness
Positive stabilisation of slide-prone slopes, slope bases or riverbanks in areas where stones are scarce. Deep-rooted shrubs and trees add extra stability to the structure.

Advantages
Locally available earth, possibly mixed with gravel stones, etc. may be used for the fill. Pre-filled bags can be used if immediate measures are necessary to prevent serious damage. Limitless possibilities to shape the structures to fit local conditions. Long life at low altitudes.

Disadvantages
Limited height.

Costs
As a rule, lower than comparable other structures.

Besides these basic variants, special methods have been developed which involve the mixing of synthetic fibres within the structures. One such method, traded under the name Texsol, can be used for the stabilisation of very steep and narrow fills between structures and along highway cuttings. Another method, developed in Switzerland under the name of Textomur, (Fig. 3.23), combines welded steel wire mesh with non woven geotextiles in the construction of earth structures (Müller, 1986); the simultaneous use of live vegetation materials is an important component of the system in order to achieve durability.

3.3.5 Vegetated crib walls (Fig. 3.24; Plates 36–41)

Crib walls have been used for a long time for the stabilisation of embankments and cuttings. They are single or double header structures formed from multiple courses of concrete, wood, metal or synthetic polymer components, which are filled with permeable soil. Combining the inert elements of the structure with live materials (live branches, saplings, turf, etc.), increases their effectiveness through better internal drainage and erosion control, and enhances their aesthetic appeal.

Materials
Branches of live shrubs and trees (see Table 2.3) and rooted saplings (see Table 2.4). One to five branches or plants are used per linear m, in combination with sods, turf or hydroseeding.

Ground Bioengineering Techniques 95

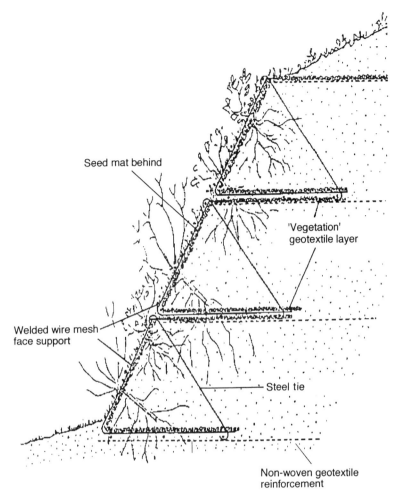

Fig. 3.23 Earth structure reinforced with wire mesh and geotextiles/geogrids. Textomur type.

Construction

Wooden poles of 100–250 mm diameter are arranged in layers as shown in Fig. 3.24 and tied together. The crosspieces or 'stretchers' of the single crib wall are anchored into the natural soil. The double wall creates a three-dimensional structure which, when filled with soil, acts as a semi-gravity wall. For the fill, loose, granular material is used which, after compaction, permits free drainage. At the same time, live branches or saplings with vigorous adventitious bud formation are placed parallel to the crosspieces. Care must be taken *not* to place these plants or branches horizontally into the soil, but at an angle of approximately 10° sloping backwards, locating the plant material so as to protrude some 250 mm

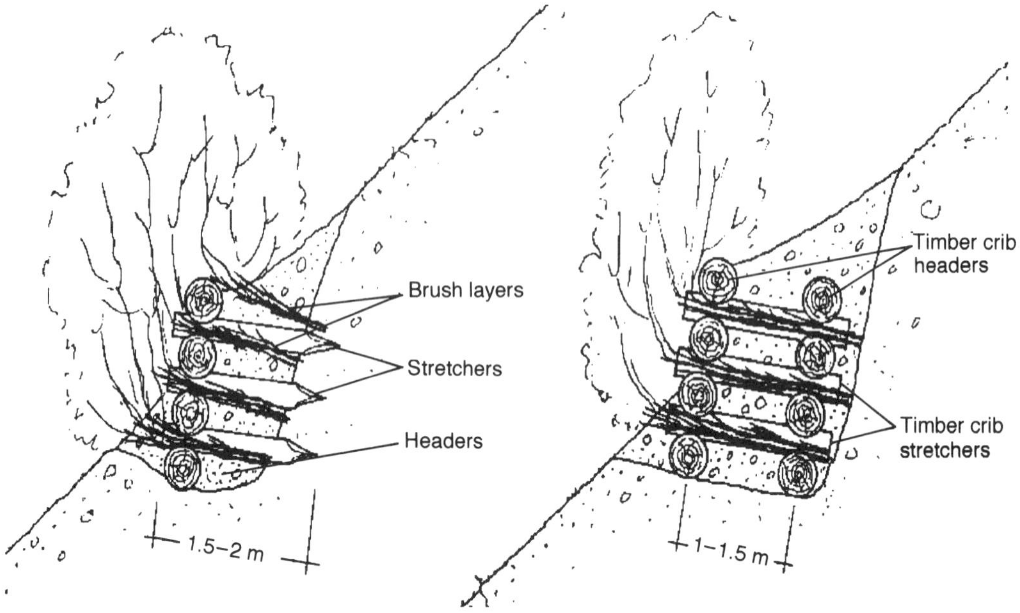

96   Ground Bioengineering Techniques

Fig. 3.24 Sectional layout of a single and double header vegetated timber crib wall.

beyond the slope surface with the cut butt ends or roots placed against the natural soil behind.

Timing
Bare-rooted saplings, stakes and live branches, only during the dormancy period; pot and container plants during the growing period; sods and turf throughout the year.

Effectiveness
The roots, stems, branches and leaf canopy of vegetation improve internal drainage, resulting in greater stability of the structure and the slope.

Advantages
Short construction period. The wooden poles may be obtained from forests nearby. The structures are very flexible.

Disadvantages
Limited height. Excessive weight of concrete component versions which are difficult to fit to the terrain, and high transport costs.

Costs
Wooden components are relatively cheap; other materials increase costs which are then similar to those of dry stone walls, stone pitching, etc.

Areas of use
Wood crib walls have been very effective during emergencies (floods in 1965 and 1966 in the Austrian Alps). Easily adapted to site conditions.

3.3.6 Live grating (Fig. 3.25; Plates 42–45)

Slope gratings consist of wood, concrete, metal or synthetic polymer elements; they require a firm foundation at the toe of the slope.

Construction
As a rule, gratings are placed on the slope, filled with free draining soil and provided with a vegetative cover. Three dimensional, i.e. double, gratings or lattice frames require large quantities of live materials, as the whole complex support structure is reinforced with live cuttings, stakes, layers of live branches, sods, or turves, etc., for extra strength. In areas where live poles of adequate length are freely available, the whole structure may be made of such local material.

Timing
If live plants and/or branches are used, only during winter dormancy. Grassing or sodding can be done during the spring/summer growing period.

Effectiveness
Live grating is effective immediately after construction is completed. After rooting of the live plant material, its stabilising and soil binding action is considerably increased. Plant water use improves internal drainage.

Advantages
Many possibilities of combining with various vegetative techniques. Immediate effect. Easily adaptable to suit slope geometry.

Areas of use
Protection and rehabilitation of slopes which cannot be flattened. Maximum height not to exceed 15–20 m.

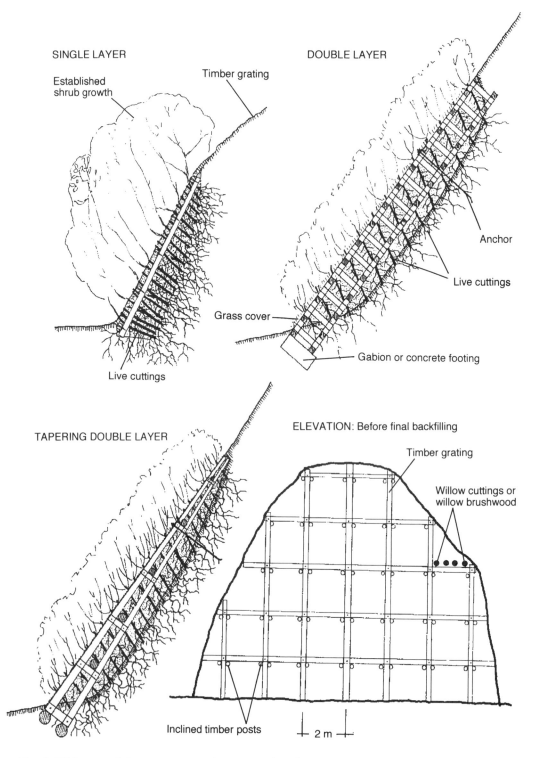

Fig. 3.25 Layout of the construction of various timber slope gratings.

3.4 Supplementary construction techniques (Fig. 3.26; Plates 47–49)

It is the aim of the supplementary plantings to increase the diversity of the primary vegetative cover and to provide for its vigour and continuity in order to achieve the objective.

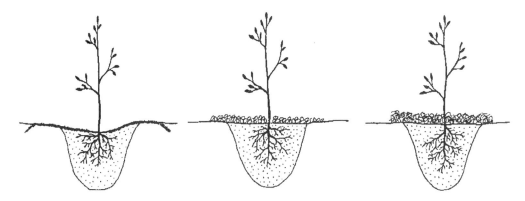

Fig. 3.26 Pit planting of saplings using thin polymer sheet or woven geotextile mulch (*left*), bark (*centre*) and straw mulch (*right*).

3.4.1 Planting of root-ball container or pot plants

The planting of bare-rooted shrubs on civil engineering sites is rarely successful because the local conditions are, more often than not, rather extreme. Such seedlings and saplings are frequently used for afforestation purposes under normal woodland conditions and they should therefore only be used where soil and climate are favourable (and as described elsewhere in this book). Pot plants and container plants with an undisturbed root system are more easily transplanted, and the failure rate is considerably lower as compared with the transplantation of bare-rooted seedings, making their use more economical. Raising seedlings in containers, pots or tubes is not only confined to shrubs and trees, but also successful in the propagation of grasses and herbs, when direct seeding methods at the construction site are unlikely to work.

Implementation
Container plants are transplanted into prepared planting holes or small pits. Depending upon location and type of plant, planting holes are prepared by hand or mechanically.

Output
A two-man team can plant between 700 and 1200 seedlings a day. Large stones have to be removed from the planting hole, to be replaced by topsoil or topsoil–manure mix. To accelerate growth, it is advisable to remove all other plants from the immediate vicinity of the seedling and to keep this area free from weeds. Different pre-formed covers or blankets can be used as mulch-mimics – plastic foil or synthetic geotextile non-wovens – or alternatively, organic mulches – bark, straw or biodegradable coir, jute or hemp geotextile mats, or by covering the soil with a thin polymer sheet or woven geotextile to eliminate competition. It is best not to use chemical weed control for this purpose. Watering the seedlings for a limited period after planting would be of great benefit by promoting early and vigorous growth.

Timing
During the dormancy period, or preferably at its transition to the growing period. In areas with a humid climate and where the transport of the planting material is rather short, transplanting can take place throughout the growing period.

Costs
Depending upon the type of plants used and site conditions, costs may vary widely, and need to be assessed for each project.

3.4.2 Transplants (Plate 49)

This involves the actual transplantation of vegetation mats, or sods up to several square metres in size, complete with topsoil and plant cover (Horstmann & Schiechtl, 1979).

Materials
Sections or panels of topsoil of the largest possible size complete with vegetative cover; plant cover of such sections may consist of grasses, shrubs and herbs, small shrubs and even tree seedlings.

Construction
The soil panels are usually cut or peeled off using scoops, back-hoes, grabs or similar mechanical equipment, and transported to the construction site by front-end loaders.

Timing
For best results, during the dormancy period.

Effectiveness
The transplantation of whole plant associations complete with roots and soil creates special growth centres on extensive construction sites prepared for re-vegetation. These islands serve as focal points for the spread of natural flora and fauna into the surrounding areas.

Costs
Only economical if suitable land for the storage of the panels is very near to the construction site.

Areas of use
Primarily for the rehabilitation of large, exposed areas if suitable topsoil sections become available as work progresses on the same project, and where suitable mechanical equipment for collection and reinstatement is locally available. (Open pit mining, quarries, borrow areas, landfills, etc.)

3.4.3 Root divisions

Materials
All plants, i.e. grasses, herbs and shrubs and trees with large and spreading root systems which permit separation without harming the plants, (e.g. *Brachypodium* spp., *Achillea millefolium*, *Ligustrum vulgare*). Of the commercially available grasses, 1 sq m of cultivated turf is sufficient for the planting of 3–15 sq m of bare ground depending upon soil conditions, temperature and gradient. The figures quoted by the suppliers are usually optimistic and only achieved at favourable sites.

Construction
Suitable plants are dug up at their natural habitat or obtained from nurseries and divided into sections which are large enough to ensure vigorous growth at transplanting. The individual plant divisions are best planted with the addition of some good topsoil and manure to the planting hole. If only short grasses are to be used in relatively dry areas, there is no need to add topsoil.

3.4.4 Transplanting rhizomes and chopped rhizomes

Materials
Live rhizomes of suitable plants (e.g. *Cynodon dactylon*, *Petasites* spp., *Hyppophae rhamnoides*). The length of the rhizomes and plant spacing

depend upon the species and the cover intensity. In general, 3–5 pieces per sq m suffice.

Construction
Individual pieces of rhizomes are placed into shallow furrows and lightly covered with earth. Stick-like rhizomes may be planted like cuttings by placing them vertically into the soil, with only a small part showing above the soil surface. On poor, stony and gravelly sites, some topsoil or compost should be worked into the soil surface before planting.

Timing
Propagation by rhizome cuttings is governed by internal growth rhythms which are specific to the type of plant; they are, however, not as marked as those of tree cuttings. Rhizomes planted during the dormancy period have the best chance of success, and it is important that they are used as soon as possible after collection. They can be kept only a very short time if stored under cool conditions in moist sand.

Advantages
Simple method to establish fast-growing plants for which seed is commercially not available.

Costs
Depending upon method of collection, species, plant spacing and site conditions, the costs are relatively high.

3.5 Special structures and techniques

3.5.1 Rockfall protection

3.5.1.1 Catch walls or barriers

Falling stones, etc. may be trapped behind special protection structures; the height of such structures depends upon the height of the rock face or scarp, ranging from low fences to substantial barriers made of sturdy plants. Rolling stones may be contained by nets, usually made of steel cable which are stretched between pairs of cables securely anchored to the slope face. Their advantage lies in their elasticity, but the large mesh size offers no protection from smaller stones.

3.5.1.2 Suspended wire mesh (Fig. 3.27)

Each length of mesh is firmly anchored above the slope face, hanging free over the rock face and, if necessary, kept under tension by tying concrete blocks to the bottom strands. Individual lengths of mesh may be stitched together with wire. Suspended protection nets are particularly suitable for very steep slopes and exposed faces of fragmented rock or soil. If feasible, the slopes are hydroseeded after the nets are fixed into place, thus stabilising the loose surface in time. Dislodged stones and boulders collect in hollows under the net, and may have to be removed on a yearly basis.

Fig. 3.27 Free hanging wire mesh protecting hydroseeded rock face.

Table 3.3 Soil protection techniques.

Technique	Areas of use	Suitability and effectiveness
Turf	Protection of erodible surfaces	1 At all sites where natural turf is available
Rolled turf*, cut turf	Riverbanks, gentle slopes landscaping	2
Grass seeding		
(a) Hayseed seeding*	At high elevation, in nature reserves, combined with other methods	2–3 In combination with other methods
(b) Standard seeding*	To topsoil only, permanent or green crop	1 For topsoil, no erosion hazard
(c) Hydroseeding**	Mechanical seeding of steep slopes, subsoil	2 Shaded areas, humid climate
(d) Mulch seeding**	Large area slopes of cuts or embankments, on subsoil or extreme sites	1 All sites, on subsoil
Seeding of shrubs and trees	To establish woodland shrubs; to supplement other vegetative methods	1 Stony, rocky, extreme slopes
Seeding over erosion control netting	Very steep slopes, sandy slopes, riverbanks	2 Erosion control
Placing of rolled turf	Waterways, even, gentle slopes	1–2
Precast concrete cellular blocks	Parking areas, access roads, protection of low embankments	2–3
Live brush mat	Protection of slopes that are exposed to water and wind erosion	1–2

Suitability: 1, very suitable; 2, suitable; 3, limited suitability

Advantages	Disadvantages	Timing	Costs
Site-adapted vegetation, immediately effective, rapid, easy application	Difficult to obtain	Growing period	Machinery: low; by hand: medium
Immediately effective	Topsoil required	Growing period	Low to medium
Site adapted, multi-species mixes	Difficult to obtain, seeding onto topsoil	Growing period	Low
Fast, low cost seeding	Presence of fertile topsoil essential	Growing period	Low
Rapid establishment, mechanisation feasible, all components applied in one operation	Vehicular access necessary, limited reach of machinery, limited success on dry, sunny slopes	Growing period	Low to average
Optimum eco-climate effects fast germination, high germination percentage, formation of an organic layer, mechanical protection	Several separate operations, on high elevation sites, slow decomposition of cover layer	Growing period	Average
Economical, natural, use in areas not suitable for grassing etc.	Slow development	Beginning or end of growing period	Low
Immediate effect	Costly	Growing period	High
Immediate effect	Fine seed bed required, preferable on topsoil	Growing period	Average to high
Immediate effect, grass establishment when in use	High work input, limited height, desiccation	Growing period	High
Immediately effective, fast growing, dense bush cover	High material input	Dormancy period	Average

*, on topsoil; **, subsoil or inert material.

Table 3.4 Ground stabilisation techniques.

Technique	Areas of use	Suitability: ecologically	technically
Live cuttings	Protection and reinforcement of slopes and dry stone walls	2–1	2
Wattle fence	Protection and retention of topsoils	2	3
Fascine	Topsoil retention, stabilisation of deep soil on wet slopes	2	2
Fascine drain	Slope drainage	1–2	2
Furrow planting	Stabilisation of wet and moderately steep slopes	1–2	2
Cordon construction (Praxl, 1961)	Stabilisation of steep slopes of heavy textured soil	2–1	2–1
Layering (a) Brush layer	Protection and stabilisation of embankments and cuttings	2	1
(b) Hedge layer	Protection and stabilisation of embankments and cuttings	2	1
(c) Hedge–brush layer	Protection and stabilisation of embankments and cuttings	1	1
Gully control	Rehabilitation of gullies	2	2
Stake fence	Rehabilitation of deep and narrow gullies	2	2

Suitability: 1, very good; 2, good; 3, limited.

Advantages	Disadvantages	Timing	Costs
Quick, simple construction, can be done after completion of works	None	Dormancy period	Very low
Immediately effective	High material input, little effect below surface, difficult rooting, sensitive to rockfall	Dormancy period	Average to high
Combined effect of drainage and water retention, simple construction	Only slender cuttings and branches can be used, sensitive to rockfall	Dormancy period	Average
Simple, quick construction	Only long branches and cuttings to be used	Dormancy period	Low to high
Combined effect of drainage and water conservation	Labour-intensive, only for gentle slopes, not for stony or rocky slopes	Dormancy period	Average to high
Effective stabilisation	High material demand	Dormancy period	Very high
Simple mechanised construction, effective at subsoil level, all shapes of branches of use	None	Dormancy period	Low
Immediately effective pioneer and climax vegetation, established at end of construction	Fertile topsoil essential, high demand for rooted plants	Growing period	Low to average
Pioneer and climax vegetation established in one operation, effective in lower soil levels	None	Dormancy period	Low
Long lasting effect	High demand for live branches	Dormancy period	Average
Rapid construction, immediately effective	Limited width and height, only for lower elevation with fine textured soils	Dormancy period	Average

Table 3.5 Combined construction techniques.

Technique	Areas of use	Suitability: ecologically	technically
Dry stone walls	Protection of slope base	1–2	2–1
Filter wedge	Stabilisation of slopes prone to slides	1–2	1–2
Gabion walls	Protection of slope base	3*	2*
Geotextile walls	Steep slopes in limited areas, protection of slopes prone to slides	2*	2*
Crib walls (a) Timber	Stabilisation of slope base, underpinning potential slides	1	2
(b) Concrete	As above	2	1
Live grating	Protection of steep slopes	1–2	2–3
Wire mesh	Stabilisation of slides at upper reaches of slope	1–2	1–2

Suitability: 1, very good; 2, good; 3, limited;

Advantages	Disadvantages	Timing	Costs
Simple construction, water permeable, flexible, limited movement possible, easy to repair	Limited height	Cuttings: dormancy period; grassing: growing period	Average
Economical	None	Cuttings, branches: dormancy period; seeding: growing period	Low to average
Quick, simple construction, permeable, flexible	Difficult to establish vegetative cover after completion, stones must be available	Dormancy period	Average
Immediately effective, elastic, adaptable to site conditions, use of local materials	Ultra-violet degradation, limited height	Dormancy period	Low to average
Locally available material, short construction time, permeable, flexible	Limited height	Dormancy period	Low
Short construction time, permeable, long life, suitable for high walls, great strength	High mass, transport, limited adaptability to site	Dormancy period	Average to high
Immediately effective, easily adaptable	None	Cuttings, live branches: dormancy period; grassing: growing period	Average
Easily adapted to site	Difficult to obtain	Depending upon live material	Average

*, estimate, as little empirical data available.

3.5.1.3 Fixed protection nets

The wire mesh is attached to steel cables which in turn are fixed to solid rock anchors. All parts of the protection net must be strong enough to cope with the forces exerted by any rockfall, in addition to snow and ice pressure; they must also be protected against corrosion, and the distance from the slope surface to the net should not exceed 300 mm. The overlap between the net sections should be 400 mm, and these should be securely tied together with galvanised wire. The lowest part of the net is left hanging free to facilitate the removal of accumulated stones and rock. Fixed protection nets on rather flat slopes should be placed as closely as possible to the ground surface, thus preventing all larger rock fragments from becoming dislodged. This requires at least one anchor point per square metre. The nets must fit snugly to the slope surface, which should be seeded as soon as possible to enhance permanent stability.

3.5.2 Wind breaks or shelters

Wind breaks or shelters fulfil several functions. The choice of materials for their construction will therefore depend upon the desired effect to be achieved. The primary aim is to reduce wind speed and, with it, its erosive force, limiting the translocation of sand and snow, or depositing it at the downwind sides of the shelter structures at pre-selected sites. A side-effect of such structures is a reduction in the rate of evapotranspiration.

Wind breaks aim to achieve the following:

- protection of agricultural land (arable land, pastures, orchards and vineyards);
- protection of afforestation areas and plantations;
- protection of buildings and structures (highways, housing, factories);
- prevention of wind erosion (dunes and sandy beaches, mine and industrial dumps, landfills);
- planned formation of sand or snow drifts.

To achieve these objectives the following structures and techniques are, among others, in common use:

- shrub and tree belts;
- fences;
- soil covers;

❏ incorporation of organic matter (straw) into the topsoil;
❏ soil reinforcing by chemical means;
❏ overhead spray irrigation.

3.5.3 Noise abatement structures (Figs 3.28 and 3.29)

The protection of humans from loud noise is essential and is of increasing concern to the health authorities. Densely wooded green belts of adequate width (100–200 m) with a high proportion of evergreen trees and undergrowth can afford reasonable protection.

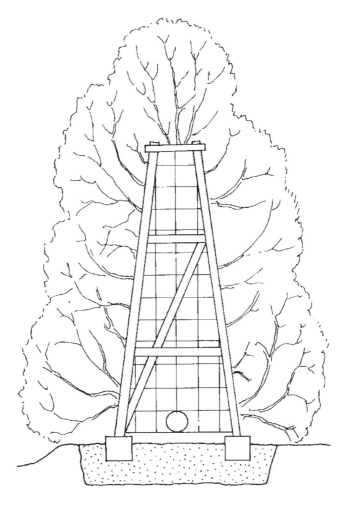

Fig. 3.28 Vegetative noise abatement barrier, Mast type.

Fig. 3.29 Noise abatement wall built of pre-cast concrete components and planted with trees and shrubs.

The shortage of available space in densely inhabited areas makes the establishment of such green belts impossible; several noise abatement systems that require little space were therefore developed. Noise abatement walls combined with vegetative cover have proved eminently suitable for the purpose, and are aesthetically pleasing to the eye. Crib walls, planted with small trees and bushes are highly effective if sufficient space is available. The best solution, however, is earth walls or bunds, grassed and planted with trees and shrubs, if space is not a problem and sufficient soil is available.

Chapter 4
Care and Maintenance of Structures

As already mentioned, it is a characteristic of bioengineering soil protection and ground stabilisation techniques that their full efficiency is only reached after the plants have rooted and active growth has commenced. To accelerate this establishment process and to shorten the time span until full effectiveness is achieved, certain care and maintenance operations must be carried out. The more extreme the local site conditions, the more intensive the care required. The costs for this are, as a rule, lower than anticipated. For example, the total maintenance costs for all vegetated areas of the German Federal road system amounts to only 5.5% of total road maintenance costs. In principle, a distinction must be made between the aftercare required during the establishment period, the upkeep during the development phase and, finally, general maintenance.

Aftercare during the establishment phase comprises all operations designed to bring the works to the required standard for commissioning, placing great emphasis on achieving vigorous growth for progressive development.

The commissioning stage is defined as follows:

- Seeding and seed mats: all areas must have an even stand of the specified grasses and herbs with not less than 50% effective ground cover. Volunteer vegetation suitable for the habitat and of equal value may be accepted as part of the ground cover.
- Seeding of shrubs and trees: seeds and mulch must be spread evenly; seedlings of shrubs and trees and those of grasses and herbs in the mix must be present in the specified proportions.
- Turves and rolled turf must be evenly rooted and firmly laid on the ground surface, i.e. with air gaps under them.
- Shrub and tree plantings: the maximum failure rate for the individual shrub or tree must not exceed 30% and the objectives must be achieved.

❑ Live plant material: fascines, brush layers, hedge layers, hedge–brush layers, wattle fences and cordons must show an average of five and a minimum of two live shoots per linear m. Live mats must show an average of ten and a minimum of five shoots per sq m, approximately evenly spaced. Two-thirds of all cuttings, live stakes, poles or truncheons must have thrown shoots, maintaining an even distribution pattern over the whole area.

These general specifications may be modified in the light of local site conditions, when they are usually made more stringent.

Aftercare during the establishment phase includes: the replacement of all plants or plant material which has not taken; cultural measures to promote plant growth such as fertilising, weeding, cultivating, mulching, irrigation; and plant protection to guard against damage by pests and disease, staking and tying, and elimination of undue competition.

Aftercare during the development phase aims at making the works fully functional. This requires on average from two to five seasons and provision for this should be made in the tender specifications. At the end of this period, all vegetative components should be at a stage that ensures their continuing progress and all works should be fully functional.

During the aftercare period of the development phase, the following cultural measures may be required:

4.1 Fertilisation

The chief purpose of the vegetation cover established during project implementation is the protection and stabilisation of earthworks, and not economic yield. Any fertiliser programme is therefore primarily aimed at enhancing the vigour of that cover. For most civil engineering projects that entail the movement of large volumes of earth, fertile topsoil is only available in limited quantities; it is therefore only occasionally unnecessary to use fertiliser to promote plant growth. It has been shown on numerous occasions that the judicious application of fertiliser to subsoil or similar material will facilitate and encourage the spontaneous establishment of plant life. A good ground cover with artificially established plants is achieved much sooner by the application of fertiliser, thus shortening the critical time period when bare soil is at risk. On poor and infertile soil, regular fertiliser applications during the first few years are essential for the well-being of the vegetation cover. The type of fertilisers used and the quantities involved will depend to a large degree upon the site conditions. There are four basic types of fertiliser

Plate 31
Filter wedge during construction

Plate 32
Gabions during construction

Plate 33
Gabions reinforced with willow cuttings after nine years

Plate 34 (top left)
Geotextile reinforced slope with branch layers

Plate 35 (top right)
Comparison with Plate 34 five months after completion: the willows are well established

Plate 36
Double stretcher log crib wall, combined with willow brush layers

Plate 37
Detail of Plate 36

Plate 38
Log crib wall after completion: the willows have not yet taken

Plate 39
Comparison with Plate 38 after five years

Plate 40
Concrete crib wall during construction: the willow brush layers are inserted at the same time

Plate 41
Concrete crib wall with willow brush layer after one year (compare with Plate 40): they reinforce and drain the fill

Plate 42
Construction of a simple wooden grating, using poles

Plate 43
Detail of a double grating using wooden poles

Plate 44
Double grating work for slope stabilisation

Plate 45
Comparison with Plate 44 after 15 years

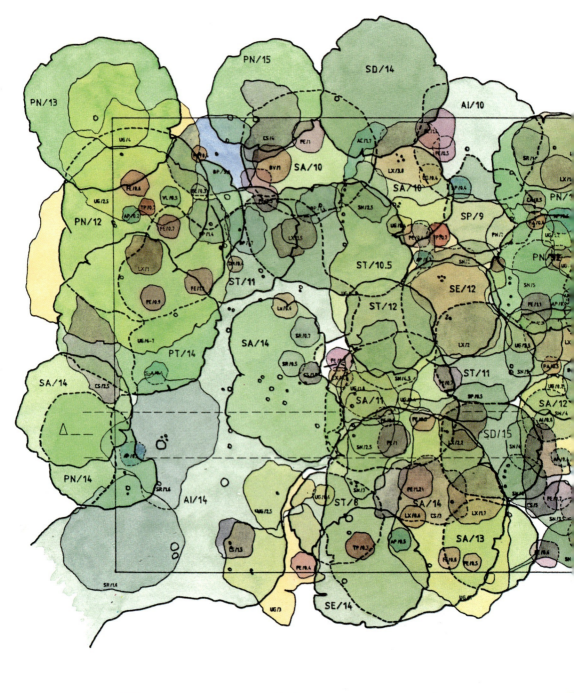

SA	Salix alba	AI	Alnus incana	UG	Ulmus glabra
ST	Salix triandra	PN	Populus nigra	TC	Tilia cordata
SP	Salix purpurea	PT	Populus tremula	TP	Tilia platyphyllos
SN	Salix nigricans	FE	Fraxinus excelsior	LX	Lonicera xylosteum
SE	Salix eleagnos	AP	Acer pseudoplatanus	PA	Picea abies

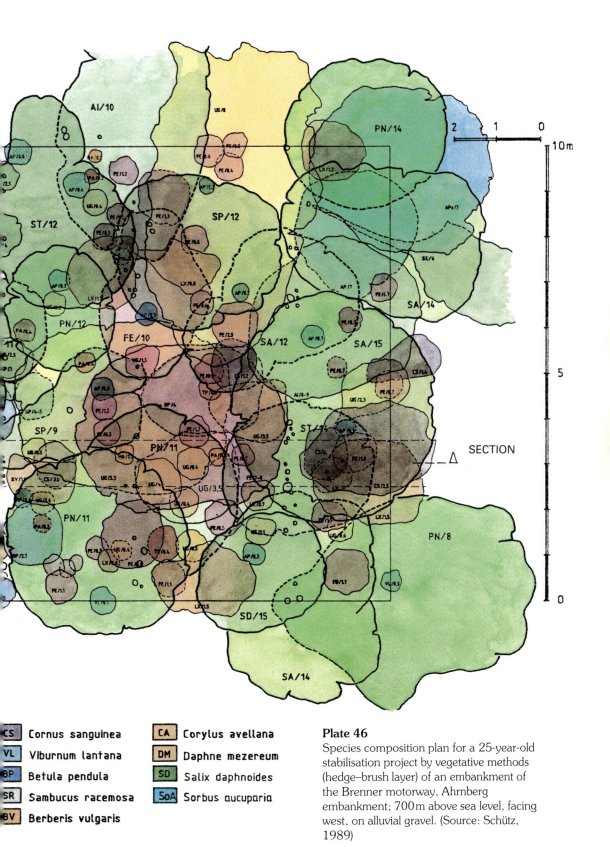

Plate 46
Species composition plan for a 25-year-old stabilisation project by vegetative methods (hedge–brush layer) of an embankment of the Brenner motorway, Ahrnberg embankment; 700m above sea level, facing west, on alluvial gravel. (Source: Schütz, 1989)

Plate 47
Slope stabilised by vegetative means before planting of woody species

Plate 48
The same slope shown in Plate 46, 15 years after planting of woody species

Plate 49
Transplantation of large turf panels complete with topsoil: small woody plants are contained in the panels

materials, each of which may be used on its own, or in combination with others:

- mineral fertiliser;
- organic fertiliser: manure;
- organic fertiliser: compost;
- green crop.

4.2 Irrigation

In regions with temperate climates, irrigation is to be used only sparingly for very short periods (at planting on dry sites, during prolonged dry spells). In humid areas, but with dry summers, occasional watering during the active growing period may be necessary. In semi-arid or arid areas, grassing and plantings without full irrigation (overhead spray) are never successful (Lecher, 1978).

4.3 Ground preparation

This consists of loosening the soil to encourage root development, and cultivation to keep the area free of weeds.

4.4 Mulching

Over the planting stations or over the whole area, organic material, such as straw which decomposes readily, is applied in a layer 100–200 mm thick. Mown grass and weeds serve the same purpose. A proper mulch modifies the temperature at ground level and regulates the moisture content of the air and topsoil horizons. This improves soil structure and plant growth (Karl, 1990). If there is the threat of a serious build-up of the rodent population, the mulch must be removed before the onset of winter.

4.5 Mowing

Mowing is not absolutely necessary; nevertheless, a single cut is recommended for all grassland. Mowing encourages plant growth and root development, and thus accelerates the formation of a dense sward of grasses and herbs, particularly if heavy seeding rates were used.

4.6 Pruning

Shrubs and trees must be pruned during the first two years after establishment for shaping and removal of dead wood, etc. Single shoots are pruned to encourage bushy growth.

4.7 Staking and tying

Saplings and larger plants are to be tied to stakes or poles, taking care not to damage the root system when pushing them into the ground. Depending upon growth rate and root development, trees need this support for a period of 3–5 years. During this time, stakes and ties are to be regularly checked and if necessary, repaired or replaced.

4.8 Pest and disease control: prevention of browsing damage by wildlife

The control of insects and fungal infection is best achieved by biological means. The most effective, but also the most expensive, protection against damage caused by game is a fence. Chemical and mechanical measures afford a certain degree of protection.

Short-term maintenance comprises all those activities which are required to keep the protection works and plantations in good order and to enhance their efficiency. This task is usually subcontracted to a firm specialising in that type of work or it is directly undertaken by the client with a team of trained personnel. If the correct vegetative method together with the most suitable planting materials were selected in the first instance, the aftercare period is unlikely to extend over more than 2–5 years.

With a successful project, the initially established plantings will develop naturally, culminating in a plant association adapted to the local conditions. This natural and self-regulating development process takes time which, in the interest of achieving early stability, should be kept as short as possible. Tree plantings in particular need, therefore, a certain amount of maintenance in the short- and in the long-term.

Care and maintenance, apart from their basic aims, may be multi-purpose:

- Permanent shrub and tree associations may serve as nurseries for live plant material. Periodic severe pruning will yield material for further use.

❏ To maintain the efficiency and proper functioning of permanent plant associations, as for example windbreaks, noise abatement belts, etc. requires periodic maintenance operations such as pruning, removal of dead wood and partial thinning.

❏ The management of protective climax woodland in potential forest areas will follow accepted sylvicultural principles. It must be borne in mind, that woodland or shrubland established for the protection of earthworks is essentially comprised of deciduous species with a very low component of conifers. Only on very rare occasions will coniferous woodland have to be dealt with. It is therefore the management principles of coppice and shrubs which are mainly applied. Judicious thinning should aim at achieving ideal stands, crown cover and understorey to ensure maximum efficiency for the intended function. On slopes prone to slides, shortened cutting cycles should aim at preventing the progression to mature forest, which on average may take 20–30 years after planting. The sylvicultural operations are to be carried out periodically every few years, preferably planned and executed by trained professionals of some forestry organisation. To illustrate what type of stand has developed after 25 years from a hedge–brush layer, plan and section details of a planned woodland association on the Ahrnberg embankment of the Brenner motorway in Tyrol, south of Innsbruck, Austria are given in Plate 46 and Fig. 4.1.

Maintenance must be carried out more often when the chosen vegetative cover is for certain reasons different to that of the natural succession. This is the case when in a potential forest area, the establishment of permanent grassland or shrubland is preferred for reasons of economic use or landscaping. Correspondingly, protracted maintenance will be necessary for grasslands and shrub associations at extreme sites, as for example on skiing pistes, mine dumps, etc. It is of advantage to prepare a maintenance plan which should guide the implementing agency during the construction period and the operating agency thereafter. If the construction period is limited to a certain part of the year, maintenance is correspondingly tied to the same period. For large construction sites, it is recommended that a maintenance plan be prepared covering the whole year to ensure that all work is carried out at the correct time, and that nothing is forgotten (see Tables 4.1 and 4.2).

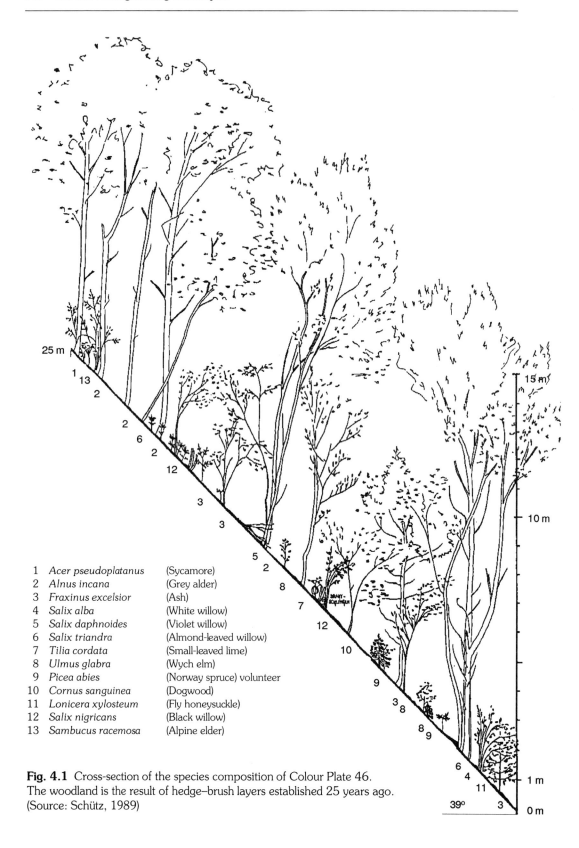

Fig. 4.1 Cross-section of the species composition of Colour Plate 46. The woodland is the result of hedge–brush layers established 25 years ago. (Source: Schütz, 1989)

1 *Acer pseudoplatanus*   (Sycamore)
2 *Alnus incana*   (Grey alder)
3 *Fraxinus excelsior*   (Ash)
4 *Salix alba*   (White willow)
5 *Salix daphnoides*   (Violet willow)
6 *Salix triandra*   (Almond-leaved willow)
7 *Tilia cordata*   (Small-leaved lime)
8 *Ulmus glabra*   (Wych elm)
9 *Picea abies*   (Norway spruce) volunteer
10 *Cornus sanguinea*   (Dogwood)
11 *Lonicera xylosteum*   (Fly honeysuckle)
12 *Salix nigricans*   (Black willow)
13 *Sambucus racemosa*   (Alpine elder)

Table 4.1 Timing of maintenance work: Temperate zone, Northern hemisphere.

Month	Type of work
March–June	Replant turf, grassland
May–September	Irrigation, soil preparation, staking and tying, fertilising, mowing, weeding, mulching, fencing
September–December	Prevention of wildlife damage
October–April	Fill-in of plantings or beating-up
December–March	Pruning: control of pioneer shrubs and trees, thinning; replacement planting of old and dead shrubs and trees
All year round	Elimination of unwanted volunteers

Table 4.2 Checklist for the care and maintenance of the vegetation used for the stabilisation and protection of earthworks.

Establishment phase – aftercare

Replacement of dead and failed plants
- Re-seeding
- Re-planting gaps
- Replacement of plants that are not thriving

Growth promotion
- Fertilising
- Irrigation
- Mulching
- Competition control: weeding, hoeing, mowing

Plant protection
- Pruning of diseased parts, sealing of cuts
- Staking and tying saplings over 1 m tall
- Fencing
- Protection against wildlife: chemical repellents, shelters and guards
- Application of fungicides and pesticides

Short-term maintenance

- Fertilising
- Mowing and grazing
- Aeration
- Maintenance pruning
- Removal of out of specification shrubs and trees
- Fence repairs
- Protection against damage by wildlife

Medium- to long-term maintenance

Forest management
- Thinning, heavy pruning
- Pruning of diseased parts
- General pruning
- Pruning to obtain fresh planting material – coppicing, pollarding

Glossary

Adventitious bud   Resting bud, specially applied bud which only shoots when required, as after damage or division

Adventitious   Formation of plant organs (shoots or/and roots) not off buds but from the permanent tissue

Adventitious   Shoots that did not develop from a normal terminal or lateral bud, or roots that did not develop from a pre-existing shoot or root

Ameliorant   Chemicals or organic matter which are applied to soils, with the intention of improving soil properties and correcting nutrient deficiencies

Anti-transpiration agents   Natural products or chemical materials which serve to avoid the drying of roots and shoots of plants during transportation or storage prior to use

Arid climates   Dry climates where the measure of evaporation exceeds the amount of rainfall

Ball-rooted tree   Tree lifted from nursery soil and transported to the planting site with the roots bare of soil

Bare-root   Seedlings or transplanted plants which have been taken directly from plant nursery beds without root ball or pot, thus are bare-rooted, and are used for reforestation and in bioengineering for species diversity

Beating up   Aftercare involving inspection of planting scheme and replacement, partially or completely, of failed plants

Bioengineering   Engineered construction method which applies biological knowledge during the project process and which uses biological materials like seeds, plants, plant parts, vegetation pieces together with inert materials to protect and stabilise slopes of:

    earthworks – embankments, cuttings and natural slope repairs – *ground* bioengineering;

    riverworks – banks of watercourses, shorelines of lakes, etc. – *water* bioengineering

Biotechnical stabilisation   Use of natural inclusions, living or inert, to reinforce soil and stabilise slopes to retain earth masses and prevent soil losses from slopes

Biotechnical slope stabilisation   Integrated or combined use of living vegetation and inert structural components

Biomass   The total mass of living and dead biological material

Biotope   Living space (locality) where organisms and their species find suitable conditions to live

Borrow areas   Sites, usually close to a project, where natural materials needed for construction, such as suitable soils, sands, gravels or rock, are excavated for use in the project and are usually abandoned or backfilled with surplus unsuitable materials at the end of the project

Brash   Small branches trimmed from the sides and top of a main stem; also known as lop and top and as slash; (v) to cut away the side branches of conifers to about 2 m height to improve access or for fire protection

Chipping   Mechanical chopping or grinding of plant materials into small particles or chips for disposal/mulching purposes

Climax community   Final stage of plant succession, which remains in its structure at that stage unless basic changes of the environment, through climatic or other influences from the outside (pasture, fire, clearing, etc.), take place; artificial climax communities that remain at a certain stage of development through occasional or continuous interferences by man (turf in potential forest areas)

Clone   Identical plant arising from a single parent by vegetative propagation

Community (plant)   Particular assemblage of plant species reflecting the prevailing environment, soil type and management

Competitive ability   Competition by an individual living organism for the limited supply of the necessities of life (e.g. space, moisture, nutrients) within its position in the food chain

Compost   A biologically very active ground improver of organic refuse (leaves, grass, shrubs, wood cuttings, bark); from landfill sites refuse composts: prior to use these must be tested for their applicability (chemicals, toxins)

Compost   Organic residues or a mixture of organic residues and soil that have been piled and allowed to undergo biological decomposition until relatively stable; compost is one of the most valuable materials for the preservation of a healthy soil and for the improvement of poor soils

Conifer   Evergreen tree that bears cones

Container plant Seeded or pot-raised plants in containers of different sizes and materials (peat, cardboard, clay, plastic); thus, independent of culture period the whole vegetation duration can be worked in; non-degradable containers are to be removed before planting

Conventional engineering (hard construction) Engineered construction incorporating inert building materials such as aggregates, rock, concrete, steel, lumber, etc. without live vegetation (apart from incidental landscape planting)

Coppice Broadleaved wood which is cut over at regular intervals to produce a number of shoots from each stool; also known as copse; (v) to cut the shoots from a stool so that more will grow

Cover crop Addition of fast-growing annual plants to a seed mix to act as a protection for the seedlings that grow more slowly; after fulfilling this task, the cover crop dies off

Crib wall A series of interconnected ladder-like bins or silos acting as mass or gravity retaining walls erected in single, double or treble rows, using alternate courses of headers (lateral members) and stretchers (longitudinal members) backfilled with free-draining granulate material for slope retaining structures – interlocking components may be logs, sawn timber, precast concrete, metal or plastic (GRP)

Cultivar Cultivated variety of a plant species, usually bred from a wild 'ecotype'

Cutting Unbranched part of a wooded stem from which, when put in the ground, a plant will develop (= live cutting); in horticulture also rhizome, culm and leaf-bud cuttings

Damp biotope Continual moist to wet ground living-space of plant and animal species; for instance lakes, springs, stream-floors, reeds, moorland, swamp/bog, meadow (fertilised), pasture

Deciduous Tree or shrub that retains its leaves for one growing season only, dropping them before the following winter

DIN standards German standards similar to those of British Standards Institute (BSI) and American Society for Testing of Materials (ASTM) but much more comprehensive for landscaping and bioengineering topics

Dripline The ground below the outermost branches of a tree's crown, where most of its feeding roots are concentrated

Ecoengineering Another term for bioengineering but without the associated overtones of medical and genetic engineering of that word

Ecology The study of how living things relate to each other and to their environment. Also used loosely to describe the interrelationship, e.g. 'the ecology of the site'

Ecological amplitude   Extent of the effect of the environmental factors (locality factors, ecological factors) for one species or species group

Ecosystem   A community of organisms interacting with one another and the environment in which they live

Ecotype   Naturally occurring variant of a species which is adapted to a particular set of ecological or environmental conditions

Erosion   Removal of surface soils and rocks by action of water, wind, frost, ice and extreme sun/heat; internal erosion leads to change of the earth structure and piping; closed vegetation is the best safeguard against erosion; extreme erosion is caused by catastrophes, e.g. floods, fires, earthquakes, etc.

Establishment   Measures which enable a possible unendangered, fast and good rooting and growth of artificially started vegetation, as for instance inoculation, composting, watering, wind and frost protection, snow cover, prevention of damage by game, fencing

Establishment period   (1) Time between sowing of the seed and the stage at which the plant is no longer reliant on the nutrient supply in the seed; (2) time between planting and the stage at which special care is not required to ensure that all parts of the plant are functioning normally

Eutrophication   Nutrient enrichments of a habitat by natural or artificial means; leads to dense and uncontrolled vegetation growth

Evapotranspiration   The whole amount of water which is taken from plant-stocks through transpiration (water evaporation over the leaves) and evaporation (water evaporation from the soil)

Evergreen   Tree or shrub that retains its leaves all year

Flora   All types of plants in certain areas

Gabion   A stone filled box or tube formed from galvanised wire mesh panels or rolls having great flexibility and strength

Geotextile   Synthetic or natural permeable fabric used in conjunction with soil and vegetation; principally for erosion control, filtration, separation, soil reinforcement and drainage

Geotextiles   Durable high tensile strength synthetic construction fabrics used for separation, filtration, drainage, reinforcement and erosion control of soils and crushed aggregates; biodegradable fabrics are made from natural fibres such as coir, jute, flax, ramie, etc. and are used primarily for erosion control, also as soil reinforcement in conjunction with brush layering (live gabions) or as short-term subsurface filters or as separators holding back soil behind geogrids in steep slopes pending establishment of vegetation

Green crop fertilisation Soil improvement by the establishment of plants which, through their root (green manure) formation and penetration, through decomposition of plant parts and leaves, or through symbiosis with fungi or bacteria, improve the nutrient content of the soil

Ground bioengineering A general term for all categories of application of plant materials for soil protection and slope stabilisation either on its own or in combination with mechanical methods

Growth habit The physical form and geometry of a plant – its above ground parts

Gully A steep sided erosion feature formed by downslope water action; unstable and recently extended drainage channel that transmits epheneral flow, has steep sides, a steep head scarp and width greater than 0.3 m, depth greater than 0.6 m; enlarges by bed scour, by head migration upslope and by side collapse

Habitat The normal abode of a plant or animal; the recognisable area or environment in which an organism normally lives

Haytrash Sweepings of seeds and broken straw from floors of hay storage barns and huts

Herb Generally non-woody flowering plant; no specific definition but covers a very wide variety of small plants; excludes grasses but merges into shrubs; also known as forb in USA

Humid climate Climate in which the yearly amount of rain is greater than the evaporation

Humus Organic fraction in the soil; decomposed plant (and animal) material

Hydroseeding The rapid application of seeds, filler, mulch, soil ameliorants and fertilisers in a water suspension onto an area where, for reasons of access, speed of application or ground condition, conventional grass seeding techniques cannot be used

Indicator plant Plant species indicating some specific characteristic of a certain locality

Inoculation Method of artificially infecting shrubs and trees or herbaceous legumes with, respectively, mycorrhizae or rhizobium bacteria that normally live in symbiosis with them

Interception Part of the moisture remaining in the vegetation layer which evaporates; depends on the meteorological conditions and the vegetation

Invaders (invading plant species) Plant species that were absent in undisturbed portions of the original plant community and will invade under disturbance or continued overuse

Lateral roots Shallow roots running out sideways from stem

Layering The development of a new individual plant from a branch or stem that has rooted in the ground

Leader The main top shoot of a tree

Legumes Herbs, shrubs and trees of the pulse family, as nitrogen gatherer because of specific root clod bacteria; good ground improvers; many form specially strong and deep-reaching roots and are good ground stabilisers; an important part of seed mixes

Light demander Plant that needs great amounts of light for optimum growth

Live-pole planting Ground bioengineering technique comprising the installation by driving, or insertion into pre-drilled holes, of long live stakes, rods or poles of 1–2.5 m length of poplar or willow (or any plant which propagates from cuttings) at close centres for slope stabilisation purposes as a form of live soil nailing

Long or pole cutting Ends of branches of tree willows and poplars, with end buds, 1–2.5 m long

Management (of vegetation) The control of vegetation for a specific purpose, to achieve a required growth habit or to manipulate the plant community

Microclimate Climate of a small area, ranging from a few square centimetres to a slope or valley bottom

Monoculture Artificial plant community (sown or planted) which is composed of a single species

Mulch, mulch layers Thick layers of plant fibres/materials, though inorganic materials can be used; in medium climatic areas, straw or hay. Originally used for the covering of ground against dehydration and erosion as suppressing weeds; today over and above as covering layer when seeding. With mechanical mulchseeding straw is being chopped and applied by way of a blower; covering layers of long-stemmed straw are spread manually or part-mechanised and tackified

Mycorrhiza A fungus which lives in symbiosis with the roots or underground parts of a plant; mycorrhizae provide plants with nutrients and often result in faster growth

Origin (provenance) Details of growth-localities or growth-space of plants generally and thickets in particular

Perennial Term for plants which grow and reproduce for many years. Perennial plants are usually woody

Pioneer species Formed of first colonising/populating pioneers; they settled on meagre nutritive substance localities and prepared these for higher successive stages; those species which are particularly well adapted to be the first to colonise bare ground

Pioneer vegetation Starting stage of the vegetation development with first-colonisation; early or young stage of a naturally developed or artificially started vegetation

Pollard A tree which is cut at 2–4 m above ground level, then allowed to grow again to produce a crop of branches; (v) to cut branches from such a tree so that they will regrow for use as poles or stakes or for purely tree management in certain areas

Position factors The whole/entirety of all external conditions (terrain factors, climatical factors, ground- and below-ground factors, human influence, animal, plant competition) which affect plants and plant species within their growing areas

Pot plants Cuttings grown in pots manufactured of various materials and of various sizes or plants transported into such pots; see also container plants

Provenance The place or origin of a tree stock, which remains the same no matter where later generations of the tree are raised. For seeding material the provenance is the harvest location; for plants it is both the harvest location and the location of the nursery

Provenance (origin) Details about location areas, about growth area and growth space of plants in general and shrubs and trees especially

Reinforced soil Mass soil structures incorporating layers of metallic, synthetic or natural materials as tensile reinforcement to facilitate construction of steep slopes and earth retaining structures

Re-naturising Restoration of alien or artificial landscape areas to the original condition

Rhizobium The nitrogen fixing bacterium capable of living in symbiotic relationship with leguminous plants in nodules on the roots

Rhizome Stem growth which creeps beneath the soil surface, rooting at nodes to form new individuals; found in many grasses and herbs

Rhizome cuttings Vegetative propagative rhizome pieces

Rill Shallow downslope erosion feature normally less than 0.3 m wide and 0.6 m deep

Riverworks Engineering works involving construction of or repairs to banks of streams, rivers, canals and edges of ponds, shorelines of lakes and sheltered portions of estuaries

Rootball plant Nursery-grown stock where roots and enclosed soil or rootballs are encased in hessian sacking/burlap to prevent root damage and desiccation through loss of soil

Root cutting Root piece which through sprouting shoots vegetatively increases

Root-characteristics

Extensive rooters Plants with a far- and deep-reaching root system. The reason for reaching deeply might be the necessity for a strong anchorage or deep ground water level (xerophytes); the penetrated ground is very often several cubic m deep; in any case, it is always many times the size of the root volume (e.g. *Petasites*, *Salix*, *Epilobium*, etc.); the food-supplying roots (haustoria) are situated either far away from the main root or very deep in the ground at the far end of the main roots; plants with taproots also belong to this group

Intensive rooters Plants with mainly dense, short-reaching and very bushy root system; these plants need large amounts of humus and may also be deep rooters if there is a good supply of nutrient substance and water in the top layer

Shallow rooters Shrubs and trees whose roots are located mainly near the soil surface

Taprooters Plants with usually a single, geotropic main root, with very few side roots of any size

Root hairs Fine structures at tips of young roots, through which water and mineral salts are absorbed from soil

Root–shoot ratio The ratio of root growth to the branches and other aerial parts of a plant

Root trainers Cylindrical ribbed plastic containers of various sizes produced in multiple lightweight packs which encourage linear root growth and thus minimise root binding, are easily prepared and transported and allow easy insertion into planting sites using planting tubes

Sapling Strong, young tree plant, the stem of which has normal branch development from the bottom up

Scrub In ecology, an area dominated by shrubs, possibly as a stage in succession to high forest; in forestry, an area of unproductive woodland

Seed bank (natural) The store of dormant seed in the soil

Seepage water Flow of water in the pores of soil under influence of gravity or capillary action

Selection Selection of plant species and varieties for a certain purpose; plants with little resilience or with high environmental requirements can be eliminated

Shrub Woody growth whose main and side shoots form multiple branches from main stock baseline or form below-ground sideshoots or on which, instead of only one stem (main stem), several stems are grown

Slip Rooted, trimmed stem of grass used for vegetative propagation

Soil bioengineering Use of plants as the major structural component for slope stabilisation, upland slope, stream banks and wetland protection systems; the cuttings serve as primary structural components, drains and barriers to earth movement

Stem cuttings Usually willow or poplar one or two years old, and straight-growing; the stems measure up to 2 m long and (long stem cuttings) are cut without side branches

Symbiosis The state of two different organisms being closely together for most/all their lives

Soil improvement Through the addition or working-in of organic ancillary materials (fertiliser, compost, green fertiliser, mulch, algae derivatives, peat, fungae, bacteria) or through the addition of dead materials (sand, rock-flour, sludge, ceramic or plastic aggregates) improvement in soil activity can be achieved

Species Group of plants similar in all respects and able to interbreed

Sprig cuttings Sprigs that can be propagated vegetatively by cutting and planting them, e.g. reed planting (*Phragmites communis*)

Stem cutting Cuttings made from shoots; according to the nature of the wood they can be divided into hardwood, semi-softwood, softwood, and herbaceous cuttings

Stolon Stem growth which creeps over the ground surface, rooting at nodes to form new individuals; found in many grasses and herbs

Stratification The use of chemical and mechanical systems to break dormancy and increase germination

Succession The process by which one community of plants gives way to another in a series from coloniser to climax; primary successions are naturally induced; secondary successions are those caused by human intervention; progressive successions are all phases (series) leading to the climax stage; regressive successions are those leading away from the climax state

Symbiosys The micro-organisms living in a larger organism (plant) in symbiosis, such as bacteria, fungi, algae

Terrace/berm A level area or bench in a section of a formed or natural slope; can be several metres wide; small terraces/berms (up to 1 m) are also called terraces; smaller terraces are termed terracettes

Thickets/copse Hedge-like formation of mostly shrub-type or bushy deciduous woods

Topsoil   Upper, rooted through, live humus surface soil layer

Transplanting   Planting of seedling in a planting bed in regular rows and intervals; according to type of plant and dimensions (height for instance) transplanting takes place more than once

Tree-felling   Also called clear-felling – total felling of a strip of forest

Turves   Cut pieces up to 1 sq m cut from natural and farmed grass. Cut turf, rolled turf; turf obtained through cutting of natural grass or specially seeded areas using machines

Vascular system   The tissues which conduct water and nutrients from one part of a plant to another – comprising xylem and phloem

Vegetation   A plant cover formed of many different plant types; the whole of the plant species of one area

Vegetation grouping   The vegetation on earth follows first of all the grouping of the climatic zones:

	Vegetation zones		
Climate	Damp	→	Dry
Cold		Tundra	
Temperate	Coniferous forest		
Medium	Decidious forest	Hardwood forest	Steppe
Sub-tropical	Laurel forest	Dry bush	Desert
Tropical	Rain forest	Monsoon forest	Savanna

Due to the large range of elevations, a variety of climates apply in the mountains and thus additional vegetation climatic sub-zones exist. Vegetation area extent, elevation and the terrain topography lead to a still finer grouping of vegetation sub-zones down to the smallest locational niche plant species

Vegetation rhythm/growing rhythm   Pattern of between growth (material production, increase) and vegetation dormancy

Vegetative propagation   Reproduction by cuttings, layering and grafting: not involving fertilisation

Wand   Usually one-year-old willow or poplar branch, up to 1 m long and up to 15 mm diameter at base, without side branches

Whip   A transplant under 1 m high

Windthrow   Wind induced instability leading to toppling of entire plant or snapping of trunk; invariably results in death of plant, though regrowth from horizontal position can sometimes occur

Woody plants   Plants containing hard tissue made up of the remains of dead xylem cells in the stems. Wood is made of mostly lignin and supports the plant and acts as a conduit for water and nutrients

Xylem Tissue in the vascular system of a plant

This glossary was prepared by Geostructures Consulting from the original text, other glossaries, including those by D.H. Gray and A.T. Leiser (1982), USDA, CIRIA (1990) and the British Trust for Conservation Volunteers, and other sources.

References

Ehrendorfer, F. (1973) Liste der Gefäßflanzen Mitteleuropas. G. Fischer, Stuttgart. (*Plant names.*)

Florineth, F. (1983) Versuche einer standortgerechten Begrünung von Erosionszonen über der Waldgrenze. Zeitschrift f. Vegetationstechnik, 11/118–122, Verlag Patzer, Berlin–Hannover. (*Soil protection methods.*)

Gray, D.H. (1991) Preface. In: *Proceedings of Workshop on Biotechnical Stabilization*, 21–23 August 1991, University of Michigan, Ann Arbor, Michigan.

Horstmann, K. und Schiechtl, H.M. (1979) Künstliche Schaffung von Ökozellen. Garten und Landschaft, 3/175–178, München. (*Transplanting of vegetation sheets or slabs.*)

Karl, S. (1990) Erfahrungen mit der Uferbepflanzung von Fließgewässern. Landschaftswasserbau, 10/425–452, TU Wien. (*Mulch layers – brush mattress.*)

Kruedener, A. (1951) Ingenieurbiologie. Verlag Ernst Reinhardt, München-Basel. (*First publication specialising in bioengineering.*)

Kuonen, V. (1983) Wald- und Güterstraßen, Eigenverlag. CH-Pfaffhausen. (*Industrial/agricultural goods and forest way construction.*)

Lecher, K. (1978) Bewässerung. Verlag Paul Parey, Berlin–Hamburg. (*Watering, irrigation.*)

Meusel, H. et al. (1965) Vergleichende Chorologie der zentraleuropäischen Flora. Verlag Fischer, Jena. (*Plant species diversity.*)

Müller, L. (1986) Geotextilien als Anker in begrünbaren Stützsystemen. TIS/4/86/182–185, Zürich. (*Geotextile retaining structures.*)

Nordin, A.R. (1993) Bioengineering to ecoengineering – Part One: the many names. In: *The International Group of Bioengineers*, Newsletter No. 3, December 1993.

Praxl, V. (1961) Der Gallinabach und sein Einzugsgebiet. Jufro – Exkursionfuehrer. Waldbau Forst. L. Bundavert. Anst. Vien 87–96. Silviculture and Forestry Federal Institute, Vienna. (*The Gallinabach*

[The Gallina Stream] and its drainage basin. Field excursion communication.)

Sauli, G. (1986) Nuove tecniche di bioingegneria naturalistica nei consolidamenti a verde possibili applicazioni nella realtà appeninica. Accad. Naz. di Scienze, XL, V, XII, I/II, 240–250, Roma.

Schaarschmidt, G. und Konecny, V. (1971) Der Einfluß von Bauweisen des Lebendverbaues auf die Standsicherheit von Böschungen. Inst. f. Verkehrswasserbau, TU Aachen. (*Slope equilibrium studies, engineering statistics.*)

Schiechtl, H.M. (1973) Sicherungsarbeiten im Landschaftsbau. Verlag G.D.W., Callwey, München. (*Fundamental reference book.*)

Schiechtl, H.M. (1980) *Bioengineering for land reclamation and conservation.* University of Alberta Press, Edmonton.

Schiechtl, H.M. (1983) Gehölze an Autobahnen. Welche sind für Dauer salzresistent? Garten und Landschaft, 11/876–882, München. (*Trees [for coppices] along motorways – which to choose for long-term salt tolerance.*)

Schiechtl, H.M. und Begemann, K. (1986) Ingenieurbiologie. Handbuch zum ökologischen Wasser-und Erdbau. (*Bioengineering handbook for ecological river- and earthworks.*) Bauverlag Wiesbaden, Auflage.

Schiechtl, H.M. (1992) Weiden für die Praxis. Verlag Patzer, Berlin–Hannover. (*Guide book for the willows suitable for bioengineering techniques in Middle Europe and the Alps.*)

Schütz, W. (1989) 25-jährige ingenieurbiologische Hangsicherungsmaßnahmen an der Brenner-Autobahn. Univ f. Bodenkultur, Wien. (*Thesis on bioengineering development and maintenance (25-year-old bioengineering slope protection measures at the Brenner Autobahn).*)

Sotir, R.B. (1995) Soil bioengineering experiences in North America. In: *Vegetation and Slopes: Stabilisation, Protection and Ecology*, (ed. D.H. Barker). Proceedings of Institution of Civil Engineers Conference, 29–30 September 1994. Thomas Telford, London.

Stärk (1963) *Slope protection.* Patent No. 230305-6, Austrian Patent Office, Vienna.

Tschermak, L. (1961) Wuchsgebietskarte des Österneichischen Waldes, Forstl. Bundesversuchsanstalt, Wien. (*Plant propagation – shrubs and small trees.*)

Further Reading

Temperate and general ground bioengineering

Abe, K. and Iwamoto, M. (1986) An evaluation of tree-root-effect on slope stability by tree root strength. In: *Journal of the Japanese Forestry Society*, **68**(12), 505–510.

Abe, K. and Iwamoto, M. (1986) Preliminary experiment on shear in soil layers with a large direct-shear apparatus. In: *Journal of the Japanese Forestry Society*, **68**(2), 61–5.

Anderson, M.G. (1984) *Prediction of soil suction for slopes in Hong Kong*. GCO Publication No. 1/84, Geotechnical Control Office, Hong Kong.

Anderson, M.G. and Pope, R.G. (1984) The incorporation of soil water physics models into geotechnical studies of landslide behaviour. *Proceedings of the 4th Symposium on Landslides*, **1**, 349–53.

Aubertin, G.M. and Kardos, L.T. (1965) Root growth through porous media under controlled conditions. I. Effect of pore size and rigidity. *Proceedings of the Soil Science Society of America*, **29**, 290–93.

Aubertin, C.M. (1971) *Nature and extent of macropores in forest soils and their influence on surface water movement*. Research Paper NE-192, Northeast Forest and Range Experiment Station, US Forest Service.

Bache, D.H. and MacAskill, I.A. (1984) *Vegetation in Civil and Landscape Engineering*. Granada, London.

Beckett, K. and Beckett, G. (1977) *Planting Native Trees and Shrubs*. Botanical Society of the British Isles.

Binns, W.O. (1980) *Trees and water*. Arboricultural Leaflet No. 6, Forestry Commission Research Station, Farnham, Surrey.

British Ecological Society (1991) *Plant root growth – An Ecological Perspective* (ed. D. Atkinson). Special Publication No. 10, Blackwell Science, Oxford.

Brown, C.B. and Sheu, M.S. (1975) *Effects of deforestation on slope*. Journal of Geotechnical Engineering Division, American Society of Civil Engineers, **101**, 147–65.

CIRIA (1990) *The Use of Vegetation in Civil Engineering.* (eds N.J. Coppin and I.G. Richards). Construction Industry Research and Information Association/Butterworths, Sevenoaks.

Crozier, M.J. (1986) *Landslides: Causes, Consequences and Environment.* Croom Helm, London.

DOT (1992) Planting, vegetation and soils. Part 2, Section 1 *The Good Roads Guide – New Roads,* Volume 10 – Environmental Design, Design Manual for Roads and Bridges. HMSO, London.

DOT (1993) The wildflower handbook. Part 1, Section 4 *Horticulture,* **10** – Environmental Design, Design Manual for Roads and Bridges. HMSO, London.

Endo, T. and Tsuruta, T. (1969) *The effect of tree roots upon the shearing strength of soil.* In: Annual Report No. 18, pp. 167–82. Hokkaido Branch, Tokyo Forest Experiment Station, Sapporo, Japan.

Forbes, P.J. (1993) *The Management of Lineside Vegetation.* Civil Engineering Department, Regional Railways, Birmingham.

Fookes, P.G., Sweeney, M., Manby, C.N.D. and Martin, R.P. (1985) Geological and geotechnical engineering aspects of low-cost roads in mountainous terrain. *Engineering Geology,* **21,** 1–152.

Gaiser, R.N. (1952) Root channels and roots in forest soils. *Proceedings of the Soil Science Society of America,* **16,** 62–5.

Gray, D.H. and Megaham, W.F. (1981) *Forest vegetation removal and slope stability in the Idaho batholith.* Research Paper INT-271, Intermountain Forest and Range Experiment Station, Ogden, Utah.

Gray, D.H. and Leiser, A.T. (1982) *Biotechnical Slope Protection and Erosion Control.* Van Nostrand Reinhold (republished by Krieger Publishing, Malabar, Florida).

Greenway, D.R. (1987) Vegetation and slope stability. In: *Slope Stability* (eds M.G. Anderson and K.S. Richards). John Wiley, Chichester.

Grime, J.P. (1979) *Plant Strategies and Vegetation Processes.* John Wiley, Chichester.

Highway Earthwork Series (1984) *Manual for Slope Protection.* Japan Road Association, Tokyo.

Holster-Jorgensen, H. (1967) Influences of forest management and drainage on groundwater fluctuation. In: *International Symposium on Forest Hydrology* (eds W.E. Sopper and H.W. Lull), pp. 325–34. Pergamon, Oxford.

ICE (1995) *Vegetation and slopes – stabilisation, protection and ecology* (ed. D.H. Barker). Proceedings of International Conference, University Museum, 29–30 September 1994, Oxford. Thomas Telford, London.

ICE (1949) *Biology and civil engineering.* Proceedings of Conference, September 1948, ICE, London.

Ishibashi, H. (1984) *Soil properties in relation to rehabilitation of the antierosion plantation. Interpraevent* 1984, Tagungspublikation, Villach.

Kitamura, Y. and Namba, S. (1981) The function of tree root upon landslides prevention presumed through the uprooting test. In: *Bulletin of the Forestry and Forest Products Research Institute* No. 313, January, Ibaraki, Japan.

Kobashi, S. (1984) The role of vegetation to slope stability. In: *Interpraevent* 1984, Tagungspublikation, Villach.

Köpke, U. (1981) A comparison of methods for measuring root growth of field crops. In: *Z. Acker-und Pflanzenbau* 150, 39–49.

Kraebel, C.J. (1936) *Erosion Control on Mountain Roads.* United States Department of Agriculture Circular No. 380, March 1936, Washington.

Lull, H.W. (1964) Ecological and soil cultural aspects. In: *Handbook of Applied Hydrology* (ed. V.T. Chow), Section 6. McGraw-Hill, New York.

Marini, F. and Paiero, P. (1988) I Salici d'Italia. Ed. Lint Trieste. (Vegetative propagatable shrubs, species selection.)

Nassif, S.H. and Wilson, E.M. (1975) The influence of slope and rain intensity on runoff and infiltration. In: *Hydrology Science Bulletin,* **20**(4), 539–53.

Nemcok, A., Pasek, I. and Rybar, I. (1972) Classification of landslides and other mass movements. In: *Rock Mechanics,* **4**, 71–78.

Oplatka, M., Diez, C., Leuzinger, Y., Palmeri, F., Dibona, L. and Frossard, P-A. (1995) *Dictionary of Soil Bioengineering.* vdf Hochschulverlag AG.

Robinson, T.W. (1958) Phreatophytes. In: *Water Supply Paper* 1423, US Geological Survey, Washington DC.

Schiechtl, H.M. (1980) *Bioengineering for Land Reclamation and Conservation.* University of Alberta Press, Alberta (republished by Permaculture Institute, Tyalgum, NSW, Australia).

Scott Russell, R. (1991) *Plant Root Systems: Their Function and Interaction With the Soil.* McGraw-Hill, Maidenhead.

Shinozaki, K., Yoda, K. and Kira, T. (1964) A quantitative analysis of plant form – the pipe model theory, I Basic Analysis. In: *Japanese Journal of Ecology,* **14**(3), 97–105.

Sidle, R.C., Pearce, A.J. and O'Loughlin, C.L. (1985) *Hillslope Stability and Land Use.* American Geophysical Union Water Resources Monograph Series.

Skempton, A.W. and Hutchinson, I. (1969) Stability of natural slopes and embankment foundations. State of the art report. In: *Proceedings of the 7th International Conference on Soil Mechanics*, Mexico, State of Art Volume.

Smoltczyk, U. (1991) *Foundation – Pocketbook*, 4th edn., Part 2. Ernst & Sohn, Berlin.

Stolzy, L.H. and Barley, K.P. (1968) Mechanical resistance encountered by roots entering compact soils. In: *Soil Science*, **105**(5), 297–301.

Strahler, A.N. (1969) *Physical Geography* (Irdedu). John Wiley, New York, p. 240.

Taylor, H.M. and Gardner, H.R. (1960) Use of wax substrates in root penetration studies. In: *Proceedings of the Soil Science Society of America*, **24**, 79–81.

Thoughton, A. (1957) *The Underground Organs of Herbage Grasses*. Agricultural Bureaux, Farnham.

Tukamoto, Y. and Kusakabe, O. (1984) Vegetative influences on debris slope occurrences on steep slopes in Japan. In: *Proceedings of the Symposium on the Effect of Forest Land Use on Erosion and Slope Stability*, Hawaii.

University of Michigan (1991) *Biotechnical stabilisation* (ed. D.H. Gray). Proceedings of Workshop, August 21–23, University of Michigan, Ann Abor, USA.

USDA (1992) Soil bioengineering for upland slope protection and erosion reduction. United States Department of Agriculture, Chapter 18 *Engineering Field Handbook* Part 650.

Van Kraayenoord, C.W.S. and Hathaway, R.L. (1986) *Plant Materials Handbook for Soil Conservation*. Vol. 1 *Principles and Practices*. National Water and Soil Conservation Authority Publication No. 93.

Viles, H. (1988) *Biogeomorphology*. Basil Blackwell, Oxford.

Vomocil, I.A., Waldron, L.J. and Chancellor, W.J. (1961) Soil tensile strength for centrifugation. In: *Proceedings of the Soil Science Society of America*, **25**, 176–80.

Waldron, L.J. (1977) Shear resistance of root permeated homogeneous and stratified soil. In: *Journal of Soil Society of America*, **41**, 843–9.

Waldron, L.J. and Dakessian, S. (1981) Soil reinforcement by roots: calculation of increased soil shear resistance from root properties. In: *Soil Science*, **132**(6), 427–35.

Welsh Development Agency (1987) *Working with nature – Low-cost Land Reclamation Techniques*. Welsh Development Agency, Cardiff.

Wu, T.H. (1984) Effect of vegetation on slope stability. In: *Transportation Research Record 965*, Transportation Research Board, Washington DC, pp. 37–46.

Yasue, T. (1984) Experimental study on the stabilization of slope by vegetation. *Interpraevent* 1984, Tagungspublikation, Villach.

Young, A. (1989) *Agroforestry for Soil Conservation*. CAB International, Wallingford and International Council for Research in Agroforestry.

German language further reading

Böll, A. und Gerber, W. (1986) Maßgebende Gesichtspunkte im Lebenverbau. Bündner Wald, 39/43–50, Chur. (*Shoots and root-growth, diseases and pests.*)

Burkhard, A.M. (1987) Festigkeitsfragen im Grünverbau. Schweizer Ingenieur und Architekt, 17/461–465, Zürich. (*Statistics, root firmness (solidity, compactness) (strength).*)

Lautenschlager, E. (1989) Die Weiden der Schweiz. Verlage Birkhäuser, Basel. (*Vegetative propagated shrubs, species selection.*)

Litzka, J. et al. (1988) Wegebau in der Landschaft, Sonderfolge von 'Der Förderungsdienst', BM f. Land- und Forstwirtschaft, Wien. (*Lane and road construction in rural areas.*)

Neumann, A. (1981) Die mitteleuropäischen Salix-Arten, Mittlgn. d. Forstl. Bundesvers.-Anstalt 137, Wien. (*Plant diversity, species choice.*)

Pflug, W. (1969) 200 Jahre Landspflege in Deutschland – ein übersicht. Deutschen Verband für Wohnungswesen, Stadtebau und Raumplanung e. V., Köln. (*200 years of land care in Germany – an overview.*) (German Association for House Design, Town Planning and Development Planning.)

Schiechtl, H.M. (1982) Anleitung zur Begrünung von Forstwegeböschungen, Österr. Forstkalender. Österr. Agrarverlag, Wien. (*Short introduction for forestry.*)

Schiechtl, H.M. (1991) Böschungssicherung mit ingenieurbiologischen Bauweisen. Grundbau-Taschenbuch, 4. Aufl. Teil 2. Ernst & Sohn, Berlin. (*Construction techniques.*)

Schlüter, U. (1986) Pflanze als Baustoff. Verlag Patzer, Berlin–Hannover. (*Application instruction book – outer alpine.*)

Schlüter, U. (1990) Laubgehölze – ingenieurbiologische Einsatzmöglichkeiten. Verlag Patzer, Berlin–Hannover. (*Species selection – lowlands – middle highland.*)

Smoltczyk, U. und Malcharek, K. (1985) Lebendverbau an Steilwänden aus Tonmergel, Jahrbuch 2 d. Gesellsch. f. Ingenieurbiologie, Bd. 2, S 170–177, Aachen. (*Geotextiles.*)

Surber, E. (1969) Nachzucht von Ballenpflanzen. Bündner Wald, 4/109–121, Chur. (*Plant growing.*)

Tobias, S. und Grubinger, H. (1988) Verbundfestigkeit – ein neuer Ansatz bei Festigkeitsfragen in der Ingenieurbiologie. Manuskript für Interpraevent 1988 in Graz. (*Bond strength – a new approach to the strength question in bioengineering.*) (Manuscript for Intrapraevent Conference 1988, Graz.)

Zeh, H. (1983) Ingenieurbiologie, Beispiele für das Bauen mit Pflanzen in der Schweiz. Garden und Landschaft, 6/471–476, München. (*Construction techniques.*)

Theses

Donat, M. (1989) Die Boden-Wurzelmatrix unter besonderer Berüksichtigung der Berasung. Univ. f. Bodenkultur, Wien. (*Slope equilibrium studies, root firmness strengthening.*)
Fritsch, S. (1984) Abmagerungsversuche bei der Rollrasenerzeugung. Univ. f. Bodenkultur, Wien. (*Turf rolls.*)
Nordin, A.B.D. Rahman (1995) *Eco-engineering practices in Malaysia.* PhD thesis, Department of Town and Country Planning, University of Newcastle-upon-Tyne.
Postl, H. (1987) Grundsätze für den naturnahen und landschaftsschonenden Wegebau. Univ. f. Bodenkultur, Wien. (*Industrial/ agricultural goods and forest road construction.*)

DIN standards and other codes of practice

DIN 18 915	Vegetation techniques in landscape construction: ground works.
DIN 18 916	Vegetation techniques in landscape construction: plant and plant works.
DIN 18 917	Vegetation techniques in landscape construction: turfing and seeding.
DIN 18 918	Vegetation techniques in landscape construction: bioengineering soil protection construction techniques by seeding, planting, construction methods with live and non-live materials and construction parts, combined construction methods.
DIN 18 919	Vegetation techniques in landscape construction: plant development and maintenance care.
Bundesamt für Strassenbau, Bern (1981)	Guidelines to natural construction including an introduction of retaining structures and noise abatement installations in the landscape.
Tschermak, L. (1961)	Plant propagation – shrubs and small trees.

Index

(The contents of Tables 2.2, 2.3 and 2.4 have not been indexed apart from the general headings, which appear in the index under broadleaved trees, conifers, grasses, herbs and shrubs.)

Achillea millefolium 101
aesthetic considerations 1–3, 6
aftercare, *table* 119
   see maintenance of structures
Agrostis stolonifera, creeping bent 8, 23
Alnus incana, grey alder 7, 34, 44
altitude
   and plant sourcing 9–10
   rooted plants suitable for, data 44–6
Anthoxanthum odoratum, sweet vernal grass 8, 24
Anthyllis vulneraria, kidney vetch 8, 30
anti-transpiration sprays 12
ash 9, 34, 44

back hoes 14–15
bark, chopped 16
Betula pendula, birch 7, 34, 44
bioengineering *see* ground bioengineering systems
biological materials *see* plants (general); broadleaved trees; conifers; grasses; herbs; shrubs
biotechnical value, attributes 8–9

birch 7, 34, 44
birds' foot trefoil 8, 30
Brachypodium spp. 101
broadleaved trees
   rooted plants suitable for vegetative construction, data 44–6
   seed availability 11
   seed suitable for vegetative construction, data 33–5
   seeding, suitable varieties 7, 33–5
   tolerant species 7
browsing damage, protection against wildlife 116–18
brush mats, erosion control 63–5
butterbur 101

Caragana arborescens, salt tolerance 9
catch walls or barriers 102
channels, grassed 50
clover 8, 32
cocksfoot 8, 30
combined masonry and vegetation construction techniques 87–98
   summary 108–9

141

composts 16
concrete cellular blocks, erosion
    control 62–3
conifers 44
    cordon construction 75–7
    rooted plants suitable for
        vegetative construction,
        data 44–6
    seeding, suitable varieties 7–8,
        32–3
    tolerant species 7
construction
    costs 19–22
        brush layers 82
        comparisons 21
        cordon construction 77
        cuttings 67
        fascine drains 73
        fascines 71
        furrow planting 75
        grass seed methods 52, 54,
            58
        layering 79
        soil protection as proportion
            of total 22
        wattle fences 69
    landscape awareness 2
    limits of applications 19
    method, selection 16–19
    rooted woody plants, *table*, 44–6
    techniques
        combined masonry and
            vegetation 20–1, 87–98
        special structures 102–20
        supplementary 99–102
    timing 19, 20
    see also ground stabilization
cordon construction 75–7
Cornus sanguinea, dogwood 7, 36,
    45
costs *see* construction, costs
couchgrass 25, 101

creeping bent 8, 23
crib walls, 94–7
currant 9, 45
cuttings 11–12, 65–7
Cynodon dactylon, couchgrass 25,
    101

Dactylis glomerata, cocksfoot 8, 30
dogwood 7, 36, 45
drainage, temporary 13
dry stone walls, vegetated 87–8
Duke of Argyll's tea-tree, salt
    tolerance 9

elder 8, 39, 46
Elaeagnus angustifolia 9, 46
    salt tolerance 9
erosion control
    concrete cellular blocks 62–3
    gullies 84–7
    live brush mats 63–5
    nets 60
    seed mats 60–1
exotics 46

false acacia 7, 46
fascine drains 71–3
fascines 70–1
fertilisation, maintenance of
    structures 114–15
Festuca rubra, red fescue 8, 26
filter wedge construction 88–90,
    108–9
Fraxinus excelsior ash 34, 44
    salt tolerance 9

gabions
    vegetated 90–1, 108–9
    free form 91
geotextile earth structures 92–4,
    108–9
glossary 121–31

grassed channels and waterways 50–1
grasses
  seed availability 10
  seed suitable for vegetative construction, data 23–9
  seeding 50–8
    dry seeding 55
    hayseeding 52–3
    hydroseeding 54–5
    mulchseeding 55–8
    standard seeding 53–4
    suitable varieties 8, 23–9
  tolerant species 7
gratings 97–8, 108–9
ground bioengineering systems
  applicability 48
  construction techniques 87–120
  definitions 5
  function and effects 5–6
  limits of application 19
  multiple effects, summary 6
  planning 1–3
  timing 19
  see also construction techniques
ground stabilising techniques 47, 65–87
  cordon construction 75–7
  cuttings 65–7
  fascine drains 71–3
  fascines 70–1
  furrow planting 74–5
  gully control 84–5
  layering 12, 77–84
  stake fences 86–7
  summary, *table* 106–7
  wattle fences 67–9
gully control 84–5

hayseeding 52–3

herbs
  seed suitable for vegetative construction, data 29–33
  seeding, suitable varieties 7–8, 23–9
Hippophöe rhamnoides, sea buckthorn 37, 45, 101
honeysuckle
  salt tolerance 9
  versatility 7, 38, 45
hydroseeding 54–5

installation schedule for bioengineering techniques 20
irrigation, maintenance of structures 115

jet, high-pressure 13–14

kidney vetch 8, 30

landscape awareness 2
  construction 2
Larix decidua, larch 7, 32, 44
layering, ground stabilisation 12, 77–84
Ligustrum vulgare, privet
  root systems 101
  versatility 7, 38, 41, 45
live gratings 97–8, 108–9
Lolium perenne, perennial ryegrass 7, 27
Lonicera tatarica, Japanese honeysuckle, salt tolerance 9
Lonicera xylosteum, fly honeysuckle 7, 9, 45
Lotus corniculatus, birds' foot trefoil 8, 30
Lycium barbarum, Duke of Argyll's tea-tree 9

maintenance of vegetative
structures 113–19
checklist 119
commissioning stage 113–14
costs 113
fertilization 114–15
irrigation 115
mowing 115–16
mulching 115
pest and disease control 116–18
pruning 116
staking and tying 116
timing 119
meadowgrass 8, 28
Melilotus albus, white melilot 8, 31
mowing, maintenance of
vegetative structures
115–16
mulching, maintenance of
vegetative structures 115
mulch seeding 55–8

nets, erosion control 60
noise abatement structures 111–12

pest and disease control 116–18
Petasites spp., butterbur 101
Pinus sylvestris, Scots pine 7, 33, 44
planning stage 1–3
checklist 3
planting techniques
pit planting 99
rhizomes 101
root divisions 101
root-ball container plants 99–100
transplants 100–1
plants for cover and construction 6–13
biotechnical value, attributes 8–9

limits of application 19
origin, and vegetation systems 9–10
propagation 10–13
salt tolerance 9
seed availability 10–11
species and planting data 23–46
broadleaved trees 33–5, 44–6
conifers 44
grasses 23–9
herbs and legumes 29–33
shrubs 35–43
see also above headings, separately
species selection 7–9
Poa pratensis, meadowgrass 8, 28
pockets 93
Populus nigra, poplar 7, 40, 44
power station projects, soil
protection techniques,
costs as proportion of total 22
preliminary works 13–19
privet, versatility 7, 38, 41, 45, 101
propagation of woody plants 10–13, 40–3
protection *see* soil protection techniques
pruning 116

red fescue 8, 26
rhizomes 101
Ribes alpinum, currant 45
salt tolerance 9
roads, soil protection techniques,
costs as proportion of total 22
Robinia pseudacacia, false acacia 7, 46
rockfall protection 102–10
root divisions, planting techniques 101

grassed channels and waterways 50–1
grasses
  seed availability 10
  seed suitable for vegetative construction, data 23–9
  seeding 50–8
    dry seeding 55
    hayseeding 52–3
    hydroseeding 54–5
    mulchseeding 55–8
    standard seeding 53–4
  suitable varieties 8, 23–9
  tolerant species 7
gratings 97–8, 108–9
ground bioengineering systems
  applicability 48
  construction techniques 87–120
  definitions 5
  function and effects 5–6
  limits of application 19
  multiple effects, summary 6
  planning 1–3
  timing 19
  see also construction techniques
ground stabilising techniques 47, 65–87
  cordon construction 75–7
  cuttings 65–7
  fascine drains 71–3
  fascines 70–1
  furrow planting 74–5
  gully control 84–5
  layering 12, 77–84
  stake fences 86–7
  summary, *table* 106–7
  wattle fences 67–9
gully control 84–5

hayseeding 52–3

herbs
  seed suitable for vegetative construction, data 29–33
  seeding, suitable varieties 7–8, 23–9
Hippophöe rhamnoides, sea buckthorn 37, 45, 101
honeysuckle
  salt tolerance 9
  versatility 7, 38, 45
hydroseeding 54–5

installation schedule for bioengineering techniques 20
irrigation, maintenance of structures 115

jet, high-pressure 13–14

kidney vetch 8, 30

landscape awareness 2
  construction 2
Larix decidua, larch 7, 32, 44
layering, ground stabilisation 12, 77–84
Ligustrum vulgare, privet
  root systems 101
  versatility 7, 38, 41, 45
live gratings 97–8, 108–9
Lolium perenne, perennial ryegrass 7, 27
Lonicera tatarica, Japanese honeysuckle, salt tolerance 9
Lonicera xylosteum, fly honeysuckle 7, 9, 45
Lotus corniculatus, birds' foot trefoil 8, 30
Lycium barbarum, Duke of Argyll's tea-tree 9

maintenance of vegetative
    structures 113–19
  checklist 119
  commissioning stage 113–14
  costs 113
  fertilization 114–15
  irrigation 115
  mowing 115–16
  mulching 115
  pest and disease control 116–18
  pruning 116
  staking and tying 116
  timing 119
meadowgrass 8, 28
Melilotus albus, white melilot 8, 31
mowing, maintenance of
    vegetative structures
    115–16
mulching, maintenance of
    vegetative structures 115
mulch seeding 55–8

nets, erosion control 60
noise abatement structures 111–12

pest and disease control 116–18
Petasites spp., butterbur 101
Pinus sylvestris, Scots pine 7, 33,
    44
planning stage 1–3
  checklist 3
planting techniques
  pit planting 99
  rhizomes 101
  root divisions 101
  root-ball container plants
    99–100
  transplants 100–1
plants for cover and construction
    6–13
  biotechnical value, attributes
    8–9

  limits of application 19
  origin, and vegetation systems
    9–10
  propagation 10–13
  salt tolerance 9
  seed availability 10–11
  species and planting data 23–46
    broadleaved trees 33–5, 44–6
    conifers 44
    grasses 23–9
    herbs and legumes 29–33
    shrubs 35–43
    see also above headings,
      separately
  species selection 7–9
Poa pratensis, meadowgrass 8, 28
pockets 93
Populus nigra, poplar 7, 40, 44
power station projects, soil
    protection techniques,
    costs as proportion of total
    22
preliminary works 13–19
privet, versatility 7, 38, 41, 45, 101
propagation of woody plants
    10–13, 40–3
protection *see* soil protection
    techniques
pruning 116

red fescue 8, 26
rhizomes 101
Ribes alpinum, currant 45
  salt tolerance 9
roads, soil protection techniques,
    costs as proportion of total
    22
Robinia pseudacacia, false acacia 7,
    46
rockfall protection 102–10
root divisions, planting techniques
    101

root-ball container plants 99–100
Rosa rugosa, rose 46
   salt tolerance 9

Salix spp. *see* willow
salt tolerance 9
Sambucus nigra, elder 8, 39, 46
sandbags 92
Scots pine 7, 33, 44
sea buckthorn 37, 45, 101
seed mats, erosion control 60–1
seeding *see* grasses; shrubs
shrubs 65–7
   rooted plants suitable for vegetative construction, data 45–6
   seed suitable for vegetative construction, data 35–43
   seeding 58–60
     seed availability 11
     suitable varieties 7–8, 35–43
   see also ground stabilising techniques
siltation 87
slopes
   preliminary works 13–16, 18
   slope gratings 97–8
   stability 18
   see also ground stabilising techniques
snowberry 9, 46
soil binding 8–9
soil improvement 9
soil protection techniques 47, 48–65
   brush mats 63–5
   costs as proportion of total 22
   erosion control nets 60
   grass seeding 50–8
   parameters 17
   planning stage 2–3
   precast concrete cellular blocks 62–3

seed mats 60–1
shrubs/trees, direct seeding 58–60
stabilising effect 17
   see also ground stabilising techniques
summary, *table* 104–5
turfing 48–50
species
   data 23–46
   lists 7–9
stabilisation *see* ground stabilising techniques
staking and tying 116
straw, use in mulch seeding 55–8
sweet vernal grass 8, 24
Symphoricarpos racemosus, snowberry 46
   salt tolerance 9
synthetics, geotextile earth structures 92–4

textile earth structures 92–4
timber slope gratings 97–8
timing of construction 19, 20
topsoil reinstatement 16
topsoil removal 14–16
trees *see* broadleaved trees; conifers; ground stabilising techniques
Trifolium pratense, red clover 8, 32
Trifolium repens, white clover 8, 32
turf walls 50
turing 48–50

vegetation systems, and plant origin 9–10
vegetative construction *see* construction
vegetative propagation 10–13
   willows 40–3
vegetative protection *see* soil protection techniques

volunteers, elimination 119

walls
  crib walls 94–7
  summary 108–9
  vegetated dry stone walls 87–8
waterways, grassed 50–1
wildlife, browsing damage protection 116–18
willow
  cordon construction 75–7
  gully control 84–5
  layering 77–84
  live brush mats 63–5
rooted plants suitable for vegetative construction, data 44–6
vegetative propagation 40–3
versatility 7
wind breaks or shelters 110–11
winter sports facilities, soil protection techniques, costs as proportion of total 22
wire mesh, rockfall protection 102–10
wire mesh baskets 91